AgeTech, Cognitive Health, and Dementia

Synthesis Lectures on Assistive, Rehabilitative, and Health-Preserving Technologies

Editors
Ronald M. Baecker, *University of Toronto*
Andrew Sixsmith, *Simon Fraser University*

Advances in medicine allow us to live longer, despite the assaults on our bodies from war, environmental damage, and natural disasters. The result is that many of us survive for years or decades with increasing difficulties in tasks such as seeing, hearing, moving, planning, remembering, and communicating.

This series provides current state-of-the-art overviews of key topics in the burgeoning field of assistive technologies. We take a broad view of this field, giving attention not only to prosthetics that compensate for impaired capabilities, but to methods for rehabilitating or restoring function, as well as protective interventions that enable individuals to be healthy for longer periods of time throughout the lifespan. Our emphasis is in the role of information and communications technologies in prosthetics, rehabilitation, and disease prevention.

AgeTech, Cognitive Health, and Dementia
Andrew Sixsmith, Judith Sixsmith, Mei Lan Fang, and Becky Horst

Interactive Technologies and Autism, Second Edition
Julie A. Kientz, Gillian R. Hayes, Matthew S. Goodwin, Mirko Gelsomini, and Gregory D. Abowd

Zero-Effort Technologies: Considerations, Challenges, and Use in Health, Wellness, and Rehabilitation, Second Edition
Jennifer Boger, Victoria Young, Jesse Hoey, Tizneem Jiancaro, and Alex Mihailidis

Human Factors in Healthcare: A Field Guide to Continuous Improvement
Avi Parush, Debi Parush, and Roy Ilan

Assistive Technology Design for Intelligence Augmentation
Stefan Carmien

Body Tracking in Healthcare
Kenton O'Hara, Cecily Morrison, Abigail Sellen, Nadia Bianchi-Berthouze, and Cathy Craig

Clear Speech: Technologies that Enable the Expression and Reception of Language
Frank Rudzicz

Designed Technologies for Healthy Aging
Claudia B. Rebola

Fieldwork for Healthcare: Guidance for Investigating Human Factors in Computing Systems
Dominic Furniss, Rebecca Randell, Aisling Ann O'Kane, Svetlena Taneva, Helena Mentis, and Ann Blandford

Fieldwork for Healthcare: Case Studies Investigating Human Factors in Computing Systems
Dominic Furniss, Aisling Ann O'Kane, Rebecca Randell, Svetlena Taneva, Helena Mentis, and Ann Blandford

Interactive Technologies for Autism
Julie A. Kientz, Matthew S. Goodwin, Gillian R. Hayes, and Gregory D. Abowd

Patient-Centered Design of Cognitive Assistive Technology for Traumatic Brain Injury Telerehabilitation
Elliot Cole

Zero Effort Technologies: Considerations, Challenges, and Use in Health, Wellness, and Rehabilitation
Alex Mihailidis, Jennifer Boger, Jesse Hoey, and Tizneem Jiancaro

Design and the Digital Divide: Insights from 40 Years in Computer Support for Older and Disabled People
Alan F. Newell

AGE-WELL NCE Inc. (www.agewell-nce.ca) is Canada's Technology and Aging Network. The pan-Canadian network brings together researchers, older adults, caregivers, partner organizations, and future leaders to accelerate the delivery of technology-based solutions that make a meaningful difference in the lives of Canadians. AGE-WELL researchers are producing technologies, services, policies, and practices that improve quality of life for older adults and caregivers and generate social and economic benefits for Canada. AGE-WELL is funded through the Government of Canada's Networks of Centres of Excellence (NCE) program.

The STAR (Science and Technology for Aging Research) Institute (www.sfu.ca/starinstitute) at Simon Fraser University (SFU) is committed to supporting community-engaged research in the rapidly growing area of technology and aging. The Institute supports the development and implementation of technologies to address many of the health challenges encountered in old age, as well as addresses the social, commercial, and policy aspects of using and accessing technologies. STAR also supports the AGE-WELL network.

NCE RCE

Networks of Centres | Réseaux de centres
of **Excellence** of Canada | d'**excellence** du Canada

STAR INSTITUTE
Science and Technology for Aging Research

AgeTech, Cognitive Health, and Dementia
Andrew Sixsmith, Judith Sixsmith, Mei Lan Fang, and Becky Horst

ISBN: 978-3-031-00477-3   print
ISBN: 978-3-031-01605-9   ebook
ISBN: 978-3-031-00039-3   hardcover

DOI 10.1007/978-3-031-01605-9

A Publication in the Springer series
SYNTHESIS LECTURES ON ASSISTIVE, REHABILITATIVE, AND HEALTH-PRESERVING TECHNOLOGIES
Lecture #14
Series Editors: Ron Baecker, University of Toronto, Andrew Sixsmith, Simon Fraser University

Series ISSN   2162-7258   Print   2162-7266   Electronic

AgeTech, Cognitive Health, and Dementia

Andrew Sixsmith,
Department of Gerontology, Simon Fraser University, Vancouver, Canada

Judith Sixsmith,
School of Health Sciences, University of Dundee, Dundee, Scotland

Mei Lan Fang,
School of Health Sciences, University of Dundee, Dundee, Scotland

Becky Horst,
Department of Neuroscience, Western University, London, Canada

SYNTHESIS LECTURES ON ASSISTIVE, REHABILITATIVE, AND HEALTH-PRESERVING TECHNOLOGIES #14

ABSTRACT

This book explores the ways in which AgeTech can contribute to healthy cognitive aging and support the independence of people with dementia. Technology can play a key role in supporting the health, independence, and well-being of older adults, particularly as a response to rapid worldwide population aging. AgeTech refers to the use of technologies, such as information and communication technologies (ICTs), robotics, mobile technologies, artificial intelligence, ambient systems, and pervasive computing to drive technology-based innovation to benefit older adults. AgeTech has the potential to provide new ways of meeting the growing demands on health and social care services to support people to stay healthy and active. As such, AgeTech represents an increasingly important market sector within world economies. The book also addresses some of the research, innovation, and policy challenges that need to be resolved if technology-based products and services are to fulfill their potential and deliver real-world impacts to improve the lives of older adults and their carers, thus promoting more inclusive communities for the benefit of all.

KEYWORDS

dementia, cognitive health, cognitive impairment, aging, technology, AgeTech, co-design, user-centered methods, ethics, innovation, health care

Contents

Acknowledgments . . . . **xv**

Abbreviations and Acronyms . . . . **xvii**

1 Introduction **1**
    1.1 The Challenge 1
    1.2 Dementia and Cognitive Health 1
    1.3 AgeTech, Dementia, and Cognitive Health 2
    1.4 The Significance of AgeTech 3
    1.5 Overview of the Book 4

2 What is Cognitive Health? **7**
    2.1 Introduction 7
    2.2 Age-Related Cognitive Changes 7
    2.3 Dementia 9
        2.3.1 How Many People have Dementia? 10
        2.3.2 Care for People with Dementia 11
    2.4 Mild Cognitive Impairment 12
    2.5 Preserving Cognitive Health 12
    2.6 Major and Minor Neurocognitive Disorders 13
    2.7 Find Out More 14

3 AgeTech for Cognitive Health and Dementia **15**
    3.1 AgeTech Research and Products for Dementia and Cognitive Health 16
    3.2 AgeTech Challenges 19
    3.3 Find Out More 20

4 Supportive Homes and Communities **21**
    4.1 The Challenge 21
    4.2 What's in This Chapter? 22
    4.3 Persona and Scenario—Mike 22
    4.4 Technology for Supportive Homes and Communities 24
    4.5 Key Initiative—Hogeweyk Dementia Village 25
    4.6 Find Out More 26

5   Health Care and Health Service Delivery **27**

5.1   The Challenge 27

5.2   What's in This Chapter? 28

5.3   Persona and Scenario—Dorothy 28

5.4   Technology for Health Care and Health Service Delivery 30

5.5   Key Initiative 32

5.6   Find Out More 33

6   Autonomy and Independence **35**

6.1   The Challenge 35

6.2   What's in This Chapter? 35

6.3   Persona and Scenario—Amrita 35

6.4   Technology for Autonomy and Independence 37

6.5   Key Initiative—The INDEPENDENT Project 39

6.6   Find Out More 41

7   Mobility and Transportation **43**

7.1   The Challenge 43

7.2   What's in This Chapter? 44

7.3   Persona and Scenario—Eva 44

7.4   Technology for Mobility and Transportation 46

7.5   Key Initiative—University of Southern California 48

7.6   Find Out More 49

8   Healthy Lifestyles **51**

8.1   The Challenge 51

8.2   What's in This Chapter? 51

8.3   Persona and Scenario—Mary-Anne and Betty 51

8.4   Technology for Supporting Healthy Lifestyles 53

8.5   Key Initiative—INDUCT Project 55

8.6   Find Out More 55

9   Staying Connected **57**

9.1   The Challenge 57

9.2   What's in This Chapter? 58

9.3   Persona and Scenario—Albert 58

9.4   Technology for Staying Connected 60

9.5   Key Initiative—The Talking Mats 62

9.6   Find Out More 62

10 Financial Wellness and Employment **65**
    10.1 The Challenge 65
    10.2 What's in This Chapter? 66
    10.3 Persona and Scenario—Edith 66
    10.4 Technology for Financial Wellness and Employment 68
    10.5 Key Initiatives 69
    10.6 Find Out More 70

11 Co-Creating Technologies with People Experiencing Cognitive Decline **71**
    11.1 Introduction 71
    11.2 Key Principles in Co-creation 72
    11.3 Co-Creation: A Step-wise Process 74
        11.3.1 Step 1: Identify Stakeholders 74
        11.3.2 Step 2: Recruitment and AgeTech Appropriation: Engaging
                People Experiencing Cognitive Decline as Co-Researchers
                and Co-Creators 75
        11.3.3 Step 3: Maintaining Engagement and Involvement 75
        11.3.4 Step 4: Capacity Building and Training 76
        11.3.5 Step 5: Co-Designing and Development 76
    11.4 Approaches and Methods for Working with People with Dementia 78
        11.4.1 Methods for Co-Designing with People Experiencing Mild
                to Moderate Dementia 79
        11.4.2 Practical Initiatives that Involve People with Dementia in
                Research and Design 80
    11.5 Persona and Scenario—Dorothy 82
    11.6 Co-creation in Practice: Challenges and Limitations 83
    11.7 Summary 84
    11.8 Find Out More 85

**12 Doing Ethical Research with People with Dementia: Challenges and
Resolutions** **87**
    12.1 Mental Capacity, Informed Consent, and Assent 88
    12.2 Communication 90
    12.3 Gatekeeping 90
    12.4 Empathy and Understanding 91
    12.5 Protection and Responsibility 92
    12.6 Persona and Scenario—Edith 93
    12.7 Voice as Participants, Co-Researchers, or Partners 94

12.8   Ethics and Cognitive Health   94

12.9   Technology in the Real World   94

12.10   Conclusion   96

12.11   Find Out More   96

13   Policy, Technology, and Cognitive Health **99**

13.1   Cognitive Health is a Worldwide Challenge   99

13.2   Social Priorities   100

13.3   Economic Priorities   101

13.4   National Plans for Dementia and Cognitive Health   101

13.5   Sustainable Development Goals   102

13.6   Policy and the Role of Technology   102

13.7   Health Service Organization and Delivery   102

13.8   Focus on Prevention   103

13.9   Supporting People with Dementia   104

        13.9.1   Supporting Caregivers   104

13.10   Wider Economic Considerations   104

13.11   Ethical Issues   105

13.12   Science Policy and Research   105

13.13   Cost Effectiveness and Impact   106

13.14   Policy Next Steps   106

13.15   Find Out More   107

**14   Demonstrating Impact—Is Technology Effective in Supporting People
        with Dementia?**. **109**

14.1   Lack of Evidence to Support Technology Adoption   109

14.2   Impact of Technology on Dementia Care and Support   110

14.3   Conclusion   111

15   Commercialization and Knowledge Mobilization **113**

15.1   Persona and Scenario—Ryan   113

15.2   Some Basic Ideas   115

15.3   Value Proposition   116

15.4   Value Proposition Template   117

15.5   Example—Value Proposition for a Non-Intrusive Home
        Monitoring System   117

15.6   Techniques and Methods for Engaging with Customers   118

15.7	Understanding the Market	119
15.8	Business Model Canvas	121
15.9	Find Out More	124

16 Emerging Issues and Future Directions ... **125**
16.1	The Word "AgeTech"	125
16.2	The Role of Technology	126
16.3	The Need for Evaluation and Evidence on the Outcomes of AgeTech	127
16.4	Caregiving, "Gatekeepers," and Caregivers	127
16.5	Reflections on Covid-19	128
16.6	Conclusion	130

AgeTech Glossary ... **131**

References ... **141**

Authors' Biographies ... **165**

Acknowledgments

The authors would like to express their appreciation for the work of Juliet Neun-Hornick, the project's Special Project Coordinator. Her excellent organizational and management skills ensured the project's momentum and vision while navigating tight time constraints and myriad project details. The authors thank J. Lynn Fraser's eagle eye, thoughtful and careful editing, as well as her organizational skills that facilitated the monograph's completion.

The authors offer their gratitude to the Morgan & Claypool Publishers' publishing team, Christine Kiilerich and Diane D. Cerra, for their deep knowledge and professionalism.

The authors would like to thank the Science and Technology for Aging Research (STAR) Institute of Simon Fraser University for its administrative and financial support and the Canadian Consortium on Neurodegeneration in Aging (CCNA) and the AGE-WELL Network of Centres of Excellence for their financial support.

The authors would particularly like to thank everyone in the AGE-WELL community—Network Office, researchers, older adults and caregivers, and partners. Their passion for innovation was essential in developing and sustaining this book's creation. Thank you for your collaborative spirit.

Abbreviations and Acronyms

AARP	American Association of Retired Persons
AGE-WELL NCE	AGE-WELL Network of Centres of Excellence
AI	Artificial intelligence
CBPR	Community-based participatory research approach
CCNA	Canadian Consortium on Neurodegeneration in Aging
CHIIP	Computerized Help Information and Interaction Project
CoPILOT	Collaborative Power Mobility for an Aging Population
DFH	Dementia Friendly Home
EDIE	Educational dementia immersive experience
HER	Electronic health records
EPSRC	Engineering and Physical Sciences Research Council
GDO	Global Dementia Observatory
GP	General Practitioner
HICs	High income countries
IADLs	Instrumental activities for daily living
ICTs	Information and communication technologies
IP	Intellectual property
IT	Information technology
IoT	Internet of Things
INDUCT	Interdisciplinary Network for Dementia Using Current Technology
IRMACS	Interdisciplinary Research in the Mathematical and Computational Sciences
KITE	Keeping in Touch Everyday
KM	Mobilization
LMICs	Low- to middle-income countries
MCI	Mild cognitive impairment
mHealth	Mobile Health
NCE	Network of Centres of Excellence
NIH	National Institutes of Health
NIHR	National Institute for Health Research
PAR	Participatory action research
PIP	Product Innovation Pathway
PESTEL	Political, economic, social, technological, environmental and legal

QRS	Quantum Robotic Systems
REACH	Responsive Engagement of the Elderly Promoting Activity and Customized Healthcare
RFID	Radio refrequency identification
SDGs	Sustainable Development Goals
SIMBAC	SIMulation-Based Assessment of Cognition
SFU	Simon Fraser University
SOM	Serviceable Obtainable Market
STAR	Science and Technology for Aging Research
SWOT	Strengths, Weaknesses, Opportunities, and Threats
TAM	Total Addressable Market
TRL	Technology Readiness Levels
UK	United Kingdom
WHO	World Health Organization
WSD	Whole System Demonstrator

CHAPTER 1

Introduction

1.1 THE CHALLENGE

Technology is increasingly seen as playing a key role in supporting and enabling the health, independence, and well-being of older adults, whatever their abilities and situations (Sixsmith, Mihailidis, and Simeonov, 2017). Advances in technology are happening at the same time as the rapid aging of populations worldwide. *AgeTech* refers to the use of advanced technologies such as information and communication technologies (ICTs), robotics, mobile technologies, artificial intelligence (AI), ambient systems, and pervasive computing to drive technology-based innovation to benefit older adults (Pruchno, 2019). AgeTech puts the needs and desires of older adults first, and harnesses the potential of technology to provide new ways of meeting the growing demands on health services and to help people to stay as healthy and active as possible (Sixsmith et al., 2017). AgeTech also represents an increasingly important market sector within world economies (Kubiak, 2016).

This book focuses on the ways in which AgeTech can contribute to *healthy cognitive aging* and support the independence of *people with dementia* by addressing some of the research, innovation, and policy challenges that need to be resolved if technology-based products and services are to fulfill their potential and deliver real-world impacts. Improving the lives of older adults and their carers by promoting more inclusive communities for the benefit of all is foundational to *responsible innovation*. Responsible innovation is a concept that requires innovators to be socially responsible when developing new products and services (Voegtlin and Scherer, 2017). As the main idea is to democratize the innovation process, a key approach in this book is to better involve persons with cognitive impairment when developing AgeTech (Lubberink et al., 2017). The book also discusses how we can better prioritize ethical and societal concerns of those living with cognitive impairment at the start of and throughout the innovation process.

1.2 DEMENTIA AND COGNITIVE HEALTH

Dementia and cognitive health represent major challenges in the present era. As people age, they typically experience minor changes in cognitive functioning. Dementia and related disorders refer to progressive decline in cognitive function (memory, communication, etc.) due to an underlying disease of the brain, such as Alzheimer's disease. Dementia has a huge impact on individuals, their families, and society as a whole. Currently, there are no treatments or cures for progressive dementia, so we need to find new ways forward that will enable people with dementia and their caregivers to continue to live independently and well. This is a major opportunity for AgeTech.

Perfect Wave: Shutterstock

1.3   AGETECH, DEMENTIA, AND COGNITIVE HEALTH

AgeTech aims to disrupt, extend, expand, or re-organize existing social practices, leading to new, more agile opportunities to provide services to improve the lives of older adults, especially those with physical difficulties, mild cognitive impairment, or dementia. Well-designed technologies that take a person-centered approach to design for diverse older adults and their families can enrich and support their complex daily routines and address challenges for all involved stakeholders, offering benefits for caregivers as well as health and social care professionals, by developing technology as a tool to reduce the complexities and stresses of daily caring activities and emotions.

The use of technology may also make health-, social-, and community-based services more adaptable to the person's circumstances and needs, as well as cost effective while also helping people to remain active contributors to their community—a win-win scenario. Health-related technologies can include tele-consultation, diagnostics, and emergency response systems (Fang et al., 2018). *E-Health* (or Telehealth) allows people to connect with health services to monitor their health conditions or receive services at home. *Telecare* includes remote monitoring, home security, device controls, and emergency response systems that enhance the safety and security, independence of older adults, and their ability to stay living at home and in their community. Other technologies include mobility devices as well as devices and applications for people with sensory impairments. Over the last decade or so, increasing attention has been given to developing technologies for people living with dementia. These include technologies for safety and security, mobility aids, reminder devices, medication management systems such as pill dispensers and devices to help with medical/health monitoring, the everyday tasks of living, and social participation. AgeTech is not just about dealing with the health problems of old age. It also emphasizes the opportunities for an increasing older demographic, for example, how can AgeTech help people to maintain their cognitive health, or help people to cope with the minor cognitive changes that are a normal part of growing older. We also need to think beyond health and think about economic and social participation and engagement in an increasingly digital world, including prospects for new businesses and services. AgeTech is not just about cost saving. It should also be about improving the access, range, and quality of services, as well as promoting quality of life and enabling social participation irrespective of the health challenges of the individual.

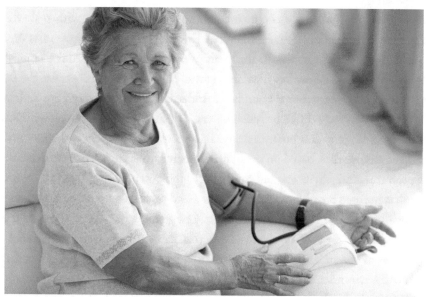

iStock.com/svetikd

1.4   THE SIGNIFICANCE OF AGETECH

AgeTech is now a part of many research and innovation programs worldwide, especially in North America, Europe, and Asia (Sixsmith, 2013). The Active Assisted Living Programme in Europe (The AAL Programme, n.d.) has invested over €700 million to develop ICT-based products and services for older adults and promote opportunities for European industry. In Canada, the AGE-WELL Network of Centres of Excellence (AGE-WELL NCE) (www.agewell-nce.ca) was launched to improve services, help people live healthy and active lives, and create new businesses in Canada. AGE-WELL NCE has propelled the AgeTech sector forward in Canada and is now globally recognized in the field. In its first five years, the network has grown to 250 researchers at 42 universities and research centers across Canada, with over 100 technologies, services, policies, and practices being developed across the network. Many of the ideas, methods, and information presented in this book are an outcome of AGE-WELL NCE's work and will be discussed later. In particular, AGE-WELL NCE defined a number of key "Challenge Areas" (see Box 3.2) that are the focus for AgeTech research (AGE-WELL NCE, 2019), such as health care, mobility and transportation, and financial wellness.

1.5   OVERVIEW OF THE BOOK

The aim of this book is to provide researchers, service providers, and entrepreneurs working in the area of technology and dementia and cognitive health with a straightforward and accessible overview of the field that will help them to combine excellent science with real-world impact. The format and style of the book is user friendly—short chapters and case studies on key topics, written in plain language. The aim is to avoid an overly academic style and technical language, while still being able to convey the key ideas and issues. This is important because the book crosses disciplinary, sectoral, and international boundaries. Indeed, we want to extend our audience beyond the academic world to include different stakeholders, especially older adults, caregivers, and health and social care practitioners. The book makes extensive use of personas and scenarios. These are semi-fictionalized case studies, based on qualitative research, that are used to illustrate the everyday experiences, problems, and situations, and to highlight the potential for AgeTech solutions.

The book is in three broad sections. Chapters 2 and 3 provide basic introductions to dementia and cognitive health, and AgeTech innovation. Chapters 4–10 look at some of the AgeTech innovations in respect to key challenge areas. Note that these are not systematic literature reviews, but brief overviews of research, technology-based products, and services. Taken together, they can be seen as a broad environmental scan of AgeTech as it relates to dementia and cognitive health. The final chapters in the book cover some of the wider issues and challenges relating to technology and cognitive health, especially challenges around ethics and working with people with dementia.

After reading this book, the reader should:

- have an understanding of the processes and outcomes of normal cognitive aging and dementia;

- appreciate the potential of technology-based products and services to meet the needs of people with dementia in key challenge areas that are important for health and well-being;

• have a better sense of the main directions for research and innovation in these challenge areas;

• understand some of the key policy and ethical challenges in the AgeTech sector;

• develop an awareness of key AgeTech products, services, and initiatives;

• identify how research can translate into real-life products and services; and

• understand good practice in working with people with cognitive impairments.

CHAPTER 2

What is Cognitive Health?

2.1 INTRODUCTION

Cognitive health concerns the ability to perform cognitive mental processes, such as learning, intuition, attention, judgment, language, and memory. Everyone's cognitive abilities tend to decline a little as we age, and we refer to this as *normal cognitive decline*. However, cognitive health can be greatly affected by age-related, debilitating disorders, such as dementia, that significantly reduce a person's quality of life and ability to live independently. Finding ways to protect and maintain the cognitive health of older adults has become a priority for many health researchers. Research has focused on developing and evaluating resources available to support those affected by cognitive disorders, while also growing the research field and mobilizing knowledge to reduce stigma, provide better preparation and prevention resources, and improve overall health. This short overview highlights the wide range of needs, situations, and issues associated with cognitive health and dementia. It is important to get a better understanding of the area before we look at the AgeTech solutions. For example, how do we maintain cognitive health? Who does dementia affect? How many people have dementia? What are the impacts on everyday life? This chapter will look at four broad categories of cognitive health in old age.

- **Age-related cognitive changes**—the minor changes in cognitive abilities as we get older.

- **Dementia**—major cognitive changes that impact on everyday life.

- **Mild cognitive impairment (MCI)**—relatively minor but noticeable cognitive changes that may be a precursor to dementia.

- **Preserving cognitive health**—the opportunities for actively maintaining and promoting cognitive well-being.

2.2 AGE-RELATED COGNITIVE CHANGES

As stated earlier, as we get older, we typically experience changes in our cognitive abilities and our ability to remember things, work out problems, and carry out tasks declines a little. For example, speed of processing information slows. Other faculties may remain intact, or even improve (such as vocabulary). These changes are referred to as *normal cognitive aging*. A small percentage of older individuals, about 1%, will continue to age without any noticeable cognitive change (Petersen, 2004), but for most older adults some degree of decline in cognitive ability will occur during the normal aging process (Albert et al., 1995; Salthouse, 2010). That subtle changes in certain cognitive processes can occur while others remain stable has been observed and is well documented in the scientific literature. Crystallized intelligence, otherwise described as the skills and abilities we have learned over the lifespan, such as vocabulary and general knowledge, remains stable through the later decades of life (Harada, Love, and Triebel, 2014). In contrast, fluid intelligence, that is one's ability to problem solve or process new information, tends to slowly decline after age 30. Fluid cognition encompasses a wide range of cognitive, such as processing speed, attention, memory, verbal fluency, visuospatial abilities, and other executive functions. While most older adults will experience the minor, age-related cognitive decline as they grow older, for most older adults this remains relatively minor with little impact on their everyday lives. The work of Rodrigues et al. (in press) highlights a number of normal, age-related cognitive changes.

- Processing speed starts to decline in the third decade of life and continues across the lifespan.

- Attention declines as we age, with the biggest impact on complex attention tasks, such as selective and divided attention.

- Memory—the ability to recall names and places is the cognitive function that most declines with age.

- Executive functioning—the capacity of a person to successfully engage in purposive behavior—declines with age, especially after age 70.

- Visual construction skills—the ability to put together individual parts to make a coherent whole (e.g., assembling furniture from parts)—declines over time.

These cognitive changes are variable between individuals and not every older adult will experience the same effects. Declines in any one of these areas, even subtly, can have an effect on the day-to-day life of an older adult. It is also important not to dismiss these minor changes, as they may be indicative of further declines in cognitive health. These declines in cognition, no matter how subtle, are critical to consider as the populations age, with the global population of people over 60 growing by about 3% per year, projecting to reach approximately 1.4 billion older adults worldwide by 2030 (United Nations, 2015). As populations continue to age, it is important to consider the number of individual older adults who will be affected by some sort of cognitive decline. Maintaining and preserving cognitive health from further age-related declines or from the effects of diseases altering cognition should continue to be a priority at all levels—individual, community, governmentally, and globally.

2.3   DEMENTIA

Beyond normal age-related changes, individuals can be severely affected by diseases that lead to symptoms that impair functions of daily living and declines in quality of life. Dementia affects the cognitive health of millions of individuals globally and is one of the main causes of disability in later life, ahead of cancer, cardiovascular disease, and stroke (Alzheimer's Association 2020). Dementia is an umbrella term that describes a group of cognitive symptoms that can be induced by pathological changes to the structure and function of the brain, such as Alzheimer's disease, stroke, or traumatic brain injury. We can distinguish between "progressive dementia," where these changes in cognitive health are insidious and progressive (e.g., due to Alzheimer's disease), and those that are potentially reversible, caused by depression, medication, alcohol, head injury, etc.

Daisy Daisy: Shutterstock

Dementia affects many cognitive processes, most notably declines in memory, attention, reasoning, and communication abilities. Early stages of progressive dementia are often overlooked as common symptoms included forgetfulness and losing track of time, which are sometimes brushed off as *senior moments* or mistaken for normal declines of aging. The middle stages of dementia, when signs and symptoms become clearer, occur when individuals or their social network most often seek help. These symptoms include declines in cognitive health related to forgetfulness of recent events and people's names, becoming lost away from home, and behavior changes (WHO, 2017a). Later stages of dementia find that memory disturbances are even more pronounced and the physical signs and symptoms are more obvious. Individuals in this stage are near total dependence on others and do not actively participate in daily activities of living such as cooking, socializing, and buying groceries (WHO, 2017b). These stages are a useful way of thinking about dementia, but things are rarely so simple in real life. The time taken to progress from mild dementia to severe dementia, and the related problems experienced, can be very different from person-to-person. A person's confusion or disorientation may also be made worse by a range of factors such as the side effects of medication, change in environment (e.g., admission to residential care), boredom, or pain.

CGN089: Shutterstock

There is no single test to determine a diagnosis of dementia. Rather, there are multiple physical examinations, imaging technology, lab tests, and observation of characteristic changes. The diagnosis of dementia is often very broad, as the specific types of dementia often overlap with their symptoms and brain changes (Alzheimer's Association, 2018). There is currently no known cure for dementia, but there are methods to help alleviate symptoms. There are a limited number of medications that target cognitive deficits, as well as cognitive and behavioral, and psychosocial interventions available (Canadian Mental Health Association, n.d.). The fact that there is no cure highlights the importance of preserving cognitive health before significant declines; it also emphasizes the importance of exploring strategies to maintain cognitive health even in stages of decline.

2.3.1   HOW MANY PEOPLE HAVE DEMENTIA?

Globally it is estimated that about 50 million people have some level of dementia, with 10 million new cases occurring each year. Within the global population, the prevalence of dementia is thought to be between 5-8%. With the growing numbers of older adults worldwide, mainly due to population aging, it is estimated that there will be 82 million people with dementia worldwide by 2030 and nearly doubling to 152 million by 2050 (WHO, 2019). Although there is debate about the causes of dementia, it is important to note that even in extreme old age, dementia is not a natural part of the aging process, and that a majority of people at these ages will not have dementia. Nevertheless, dementia is an increasing challenge to health and social care providers. Countries worldwide are currently experiencing significant population aging and the numbers of people with dementia worldwide are rising as a result.

2.3.2   CARE FOR PEOPLE WITH DEMENTIA

While the growing numbers of people with dementia highlight the scale of the issue, a further challenge is how to most effectively meet the complex, intensive, and economically expensive care needs of people with dementia. Over the years, very limited dementia care was provided and even where it did exist it was usually to be found in long-term, residential facilities, while community, home-based care was often unavailable or patchy. Services often suffered from a lack of money, poor facilities, and poor staff training (MacDonald and Cooper, 2007). At best, care was *institutional* in approach and family caregivers often had to cope with very little help and support. People with dementia and their caregivers face a wide range of problems in their everyday lives, due to their declining cognitive abilities. We will look at some of these in later chapters and illustrate them with personas and scenarios. We will also look at some of the existing and emerging good practices in health care and social and community supports. However, ideas from the social model of disability, such as Kitwood (1997), have highlighted that many of the problems faced by people with dementia are often made worse by health and social care practices, and indeed by ageist attitudes within society (Blackman et al., 2003). The psychologist Tom Kitwood highlighted some of these bad care practices (1997):

- not explaining things properly, or talking too fast for the individual to understand, or using words that are too difficult;

- doing things for someone that they can do for themselves, however clumsy and slow they are; and

- not asking seriously about the individual's feelings, especially when they are anxious, upset, and confused.

It is important that technology-based solutions do not embody these kinds of bad practices and assumptions. Increasingly, research and AgeTech development is attempting to understand the lived experiences of individuals with dementia and how to engage them as active participants in research with a *voice* to be engaged with, rather than as passive research subjects (Bartlett and Martin, 2002). We will look at this further in later chapters. Methodological development, and a better understanding of the specific health and quality of life needs of people with dementia are driving the development of more appropriate services, particularly support services enable people with dementia to live independently for as long as possible, delaying or even removing the need for admission into residential care. However, supporting people with dementia and their caregivers remains one of the biggest challenges facing health and support services in the 21st century and we will explore these in more detail in subsequent chapters.

2.4   MILD COGNITIVE IMPAIRMENT

While dementia and age-related cognitive change illustrate the different ends of cognitive health, there is a middle ground with the diagnosis of mild cognitive impairment (MCI). MCI is often referred to as a *gray zone* between normal cognitive aging and early dementia (Geda, 2012). While the definition of MCI is still vague (Fang et al., 2016), it is seen as distinct from both normal cognitive decline and dementia. Although much less severe than dementia, MCI is significantly more severe than age-related memory loss and has been associated with higher rates of further cognitive decline (Peterson, 2011). MCI may develop into progressive dementia, but this is not always the case.

As the name suggests, MCI impacts cognitive health in a more mild form. It is defined as a decline in cognition greater than what is expected for normal aging, but does not interfere with activities of daily living (Petersen et al., 1999). Although short-term memory function might be impaired, it typically does not impact an individual's ability to complete tasks like laundry, cooking, or driving. Current evidence suggests that MCI may be a transition stage between normal cognitive health and dementia (Burns and Zaudig, 2002). Individuals with MCI have been seen to develop dementia at a rate of 10–30% annually (Petersen et al., 1999). However, not all individuals with MCI progress to dementia; some remain at a stable state or can even return to normal levels of cognitive health. This phenomenon demonstrates the importance of understanding how to support and maintain cognitive health, not only for those with MCI, but all older adults.

2.5   PRESERVING COGNITIVE HEALTH

For many years, it was thought that severe cognitive decline, or dementia, was an inevitable part of growing old. However, we have already seen that most older adults do not have dementia and remain active, despite the minor changes of normal cognitive aging. In recent years there has been increasing attention of what we can do to promote cognitive health as we age—what can we do to avoid dementia? What can we do to minimize the effects of normal cognitive aging?

Rodrigues et al. (in press) argue that the term cognitive health is vague and difficult to define, and that a person cannot be deemed "cognitively healthy," solely on the absence of symptoms. Cognitive health is a long-term phenomenon that varies across an individual's lifespan. It should be noted that cognitive decline in old age is likely to have its roots earlier in life (Wang et al., 2019). The latency between brain changes and cognitive symptoms means that cognitive ill-health may also be far along before any observable signs of cognitive decline become apparent. Cognitive health is related to having a healthy brain, which in turn is related to cardiovascular, physical, and hormonal health, but there is likely to be a large variation among individuals in terms of cognitive function irrespective of brain health.

The concept of cognitive health encompasses our ability to successfully perform the wide variety of mental processes our brains are able to accomplish. We hear a lot about brain training and doing puzzles to help preserve these abilities, but maintaining cognitive health involves much more than just exercising mental capabilities. When looking at the cognitive health of older adults there is a complex interaction among cognitive health, physical health, and social health. Many of the challenge areas discussed later in this book illustrate the overlap as to how these types of health can positively influence cognitive health. Understanding the interaction among these various domains and their influence on cognitive health can assist in strategies to address the challenge area of maintaining cognitive health in aging. The Cleveland Clinic (2019) illustrates six lifestyle pillars that have a profound impact on brain health. The idea of brain health encompasses both the structural and physical health of the brain in addition to the cognitive functioning component. These six pillars include: physical exercise; food and nutrition; medical health; sleep and relaxation; social interaction; and mental fitness. The National Institute on Aging, U.S., also identifies a set risk to cognitive health, including: genetic factors; medical health; and lifestyle factors such as: lack of physical activity; poor diet; smoking; excessive alcohol intake; sleep problems; and social inactivity (National Institute on Aging, 2017). The Alzheimer's Association, of the U.S., identifies three risk factors for dementia, including lack of physical activity, poor diet, and cardiovascular health (Alzheimer's Association, 2018). Although the risk factors from each source do not fully align, there is a complexity to cognitive health that needs to be recognized when facing the challenge of maintaining cognitive health in aging. While age-related cognitive changes and indeed dementia may occur, there is increasing evidence that there are things we can do to offset the risk of dementia (Centers for Disease Control and Prevention, 2009).

2.6   MAJOR AND MINOR NEUROCOGNITIVE DISORDERS

The Diagnostic and Statistical Manual of Mental Disorders is an internationally accepted classification of mental disorders (American Psychiatric Association, 2020), with the aim of improving diagnoses, treatment, and research. The most recent version of the manual, now referred to as the DSM-5, introduced the term *Major Neurocognitive Disorder* to replace the commonly used term dementia in its revised classification (American Psychiatric Association, 2014). It highlighted different etiologies (or types), such as Major Neurocognitive Disorder due to Alzheimer's disease. The DSM-5 also introduced the term *Mild Neurocognitive Disorder* to describe noticeable, but minor changes to cognitive functioning over and beyond what is normally seen with aging. This term is based on the concept of MCI (Sachs-Ericsson and Blazer, 2015). The aim is to provide a clearer definition and because *dementia* is pejorative and perhaps for reasons relating to precise relationship to milder forms of cognitive disorders. However, the term dementia is well-known to the public, and the new DSM categories and definitions are not widely accepted as yet outside clinical practice. For this reason, we will carry on using the terms dementia and MCI in this book.

2.7   FIND OUT MORE

The National Institute of Health: Cognitive Health and Older Adults—provides information and articles on cognitive health and older adults. It can be used as a great resource to understand the disorder and learn about new findings and inventions related to cognitive health: https://www.nia.nih.gov/health/cognitive-health-and-older-adults.

HealthDay—news for healthier living is a resource for people looking for more information on Alzheimer's disease. It gives a quick definition of Alzheimer's disease, and its causes, symptoms, and treatments: http://www.healthday.com/.

Centers for Disease Control and Prevention (2009)—What is a healthy brain? New research explores perceptions of cognitive health among diverse older adults—discusses people's perceptions of cognitive health and how these are influenced by factors such as race and ethnicity, physical activity, nutrition, and media: https://www.cdc.gov/aging/pdf/perceptions_of_cog_hlth_factsheet.pdf.

CHAPTER 3

AgeTech for Cognitive Health and Dementia

There has been increasing interest over the last 10–15 years about how technologies can support cognitive health and dementia. Early research and innovation focused on technologies to monitor people to ensure their safety and security (Sixsmith, 2006), for example, to detect falls in the home (Tchalla et al., 2013). This is still an important area of research. However, other areas have emerged in recent years, including:

- technologies to aid formal and informal caregivers with their roles (Maisen et al., 2018);

- companion and caregiving robots (Moyle et al., 2017b);

- technologies that allow people to live independently, such as smart homes that help with simple tasks (Meiland et al., 2017; Van der Roest et al., 2017);

- technologies for the diagnosis and treatment of patients, such as Virtual Health Assistants to gather health information and report changes (Sixsmith, 2006);

- telemedicine and e-Health, such as virtual communication with health care professionals and personal medical recording (Sixsmith et al., 2010); and

- technologies that measure brain health (Segkouli et al., 2018).

Much of the attention has focused on people with dementia and the problems they experience. This is understandable given the very significant social and economic challenges it brings. However, less attention has been given to how technology can be used to positively contribute to improving the quality of life of individuals with dementia and support cognitive health during the aging process. For example, Sixsmith et al. (2010) looked at how technologies could be developed to allow individuals to engage with enjoyable activities such as listening to music. Also, relatively little attention has been given to individuals with MCI, older adults who are experiencing normal age-related cognitive decline, and to how technology might be used to actively promote cognitive health. Arguably, research and innovation in these areas could be particularly fruitful, as long as issues of technology usability and adoption are addressed. It is also crucial to consider the ethical issues in developing and implementing technologies.

Box 3.1 INDUCT Innovative Training Network on Technology and Dementia

INDUCT (https://www.dementiainduct.eu)—is a multidisciplinary educational research framework to improve technology and care for people with dementia, and to demonstrate how technology can improve the lives of people with dementia. INDUCT (see also Chapter 8) stands for Interdisciplinary Network for Dementia Using Current Technology and is supported through the European Union's Horizon 2020 Marie Sklodowska Curie Innovative Training Networks program. INDUCT addresses three areas: (1) technology in everyday life; (2) technology for meaningful activities; and (3) health care technology.

3.1   AGETECH RESEARCH AND PRODUCTS FOR DEMENTIA AND COGNITIVE HEALTH

Technology and innovations for supporting cognitive health of older adults, as well as caregivers, are becoming more widespread in the market. The following are examples of some of the technologies (including products, projects, services, and programs) targeted toward supporting cognitive health in various ways.

Assistive devices enable people with dementia to continue to live in their own homes and communities. These include assistive technologies that help older adults by monitoring the environment and aiding with simple tasks. A variety of products aiming to accomplish this already exist on the market, such as wayfinding technologies, GPS locators that monitor the location of individuals outdoors, and motion sensors as well as cameras to use within their homes (Bossen et al., 2015). Assistive devices include simplified items such as television remotes and telephones to make utilizing these devices easier, proximity reminders to drink water, or turn off lights before leaving a room. Apps as well as automatic or *smart* medication dispensers can facilitate medication management. These devices allow for personalized timing and reminders to take medications that can also be tracked and updated via smartphone of the user or caregiver.

Safety and security have already been mentioned and continue to be an important line of research and development. The use of home monitoring systems has also been utilized for predictive and preventative safety that monitor changes in behavior to assess for fall risks or other safety hazards associated with cognitive health declines (Williams et al., 2013). Less complex safety innovations have also made their way onto the market ranging from automatic light shut-offs, stove and tap shut-offs, to phone call blockers and timed locks. These technologies are specifically aimed toward the safety and continued home autonomy of individuals experiencing cognitive health declines, but also provide caregivers with peace of mind. Future technology developments should continue to focus on the needs and wants of the individual while balancing safety priorities.

Supporting family caregivers aims to aid the responsibilities and emotional demands of taking care of a loved one experiencing cognitive health declines (Spencer, 2017). This includes the online forums, apps, and telehealth programs that are available to connect caregivers to mental and emotional supports, either through professional counseling, peer-to-peer support, or educational resources (Arntzen, Holthe, and Jentoft, 2016; Holthe et al., 2018). In addition, such supports are extremely useful as communication technologies during times when carers are not able to leave the home, or leave the cared-for person unattended such as when infection control is necessary.

Telemedicine and telehealth have been utilized to provide virtual communication with health professionals with whom the individual has difficulty accessing in person. This communication can include a simple phone call or a video conference to visually see the individual and conduct distance examinations (Salazar, Velez, and Royall, 2014). Similar technology could be used in the form of portable diagnostic tools on a small laptop or iPad interface (Segkouli et al., 2018). This would allow health practitioners to travel easily to remote areas without carrying extensive materials or tools. Accurate cognitive assessments that can utilize telecommunication tools and strategies have important implications for providing accessible care to rural areas or individuals with limited access to transportation. Technology that assists in the efficiency and accuracy of cognitive health diagnostics is an important technology to consider. Currently, there are AI and virtual health assistant projects that are examining the effectiveness of their ability to gather data on an individual level and at large to aid in the diagnostic and treatment of cognitive health patients (Morrison et al., 2017). These can be used by professional health care practitioners to monitor the cognitive health of an individual over time and report changes.

Social and emotional connections are generally felt to be good for mental health that technology can support through virtual apps and platforms and robotics (Hodge et al., 2018; Welsh et al., 2018). The use of robotics for social connection is a fairly novel development that has started to take off in recent years. Social companion robots that mimic the presence of a pet, like a dog, cat, or baby seal, are popular for assisting a person with dementia find emotional calm and purpose (Moyle et al., 2017a). More complex robots are now being developed to replicate human interaction and conversation. These are designed to assist in decreasing the feelings of social isolation and loneliness many older adults experience, but can be modified to become a social companion as well as a telepresence for caregiver information (Vaughn, Shaw, and Molloy, 2015). In addition, when infection control is necessary, such as when flu is prevalent in a population, then these sorts of technologies can be beneficial in bridging families with their loved ones.

Gaming devices and robots can also be utilized for physical activity to promote engagement and cognitive health. Using Wii Sports in a retirement home has been investigated, as well as a robot that dances and encourages others to join in (Fujisoft Inc., 2018). The application of robots is just being realized and continued investigation of the cultural and generational perception must still be investigated (Douglass-Bonner and Potts, 2013; Smeddinck, Herrlich, and Malaka, 2015).

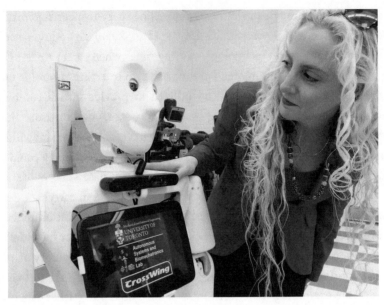

Figure 3.1: Dr. Goldie Nejat, an AGE-WELL project co-lead, is developing social robots that can prompt older adults with cognitive challenges to do activities of daily living. Photo courtesy of AGE-WELL NCE.

Cognitive stimulation through technology is an area that is growing. An increasingly large amount of games and activities to "exercise" cognitive abilities is growing. A quick look through a mobile app store can numerous find virtual puzzles, word games, visuospatial training, and a number of games that require critical thinking and other executive functions to play effectively (Kong, 2015). Games developed for therapeutic purposes are also called "serious" games.

Leisure and hobbies can be supported via technology. Digital photobooks for reminiscing have gained popularity (Karlsson et al., 2017) as has the use of tailored music through simplified music players or online streaming services. Retro-styling of such technologies can make them more familiar and perhaps more acceptable to older adults.

Electronic health records (EHR) are slowly gaining traction in many areas of the world. While paper records are still prevalent, digital health records are making collaboration across health professions and care settings more practical and efficient. This has been especially relevant for the care and support for persons with dementia, as collection of data across cohorts and clinical databases provides information from real-world patients concerning disease progression and management (National Academies of Sciences, Engineering, and Medicine, 2016).

The Internet of Things (IoT) is the concept that multiple devices or *smart devices* can be interconnected to communicate via the Internet both with other devices and the user (Dey et al., 2018). Some examples of these devices include: voice-activated assistants; fitness trackers; smart thermostats; media systems; cleaning devices; and even refrigerators. While this market is undeveloped and there is a growing subset deemed the "Internet of Medical Things," specifically focusing on the development and interconnection of: medical-grade and health-centered devices; individuals; caregivers; and professional health care providers (Gatouillat et al., 2018). There have already been important implications and technological developments targeted toward the care of dementia.

3.2   AGETECH CHALLENGES

In this book we will look in more depth at how technology can be used to address a number of challenge areas that are key areas in AgeTech research and innovation. As stated in Chapter 1, a challenge is an important, but difficult and complex problem area (sometimes known as a *wicked problem*) that demands innovation and the application of real-world solutions. A challenge may be about making positive contributions to societies, government policies, and economies. Additional challenges in this field relate to how these systems are designed and developed and how they are perceived, appropriated, and integrated into domestic and institutionalized care settings to provide positive contributions. There is more to a challenge than just ordinary research questions or priorities. For example, a further challenge concerns the identification of factors for successful and sustainable implementation in the multifaceted and challenging everyday life of a person with dementia and their social and care network. In the context of older adults, particularly those with cognitive deficits, a challenge is engaging and is worthwhile to pursue because it:

- results in significant social and economic benefits to older adults;

- is difficult to accomplish but is potentially solvable;

- requires collaboration across many disciplines and groups;

- pushes scientific understanding, knowledge, and application;

- captures popular imagination and political support;

- inspires hope; and

- brings people together to work for the common good.

Box 3.2 AGE-WELL NCE's Challenge Areas

A number of AgeTech challenge areas were identified by AGE-WELL NCE in Canada and were an outcome of consultation with over a thousand stakeholders (older adults, caregivers, service providers, and industry).

1 Supportive Homes & Communities
2 Health Care & Health Service Delivery
3 Autonomy & Independence
4 Cognitive Health & Dementia
5 Mobility & Transportation
6 Healthy Lifestyles & Wellness
7 Staying Connected
8 Financial Wellness & Employment

Courtesy of AGE-WELL NCE

The following chapters will examine these challenge areas in more detail with a specific focus on cognitive health and dementia. Each chapter will provide an overview of the challenge area, together with examples of research, products, and services that have been developed to meet these challenges. Personas and scenarios will be used to illustrate the real-world challenges that people face.

3.3   FIND OUT MORE

Information on the INDUCT network can be found at: https://www.dementiainduct.eu.

Information on research being carried out by the AGE-WELL Network can be found at: https://agewell-nce.ca/research/research-themes-and-projects.

AGE-WELL NCE's eight challenge areas—https://agewell-nce.ca/wp-content/uploads/2018/09/AGE-WELL-Challenge-Areas.pdf.

The McMaster Healthy Aging Portal provides trustworthy information on health aging based around AGE-WELL NCE's challenge areas—https://www.mcmasteroptimalaging.org.

A useful review of of assistive technologies for people with dementia is provided by Van der Roest et al. (2017).

CHAPTER 4

Supportive Homes and Communities

4.1 THE CHALLENGE

Both researchers and policy makers have identified the notion of aging in place as a driver to promote the abilities of and conditions for older adults to live independently in their own home and community (Sixsmith et al., 2017). Helping older adults to age in place, perhaps via age-friendly design, can benefit their quality of life and also provide a more appropriate and cost-effective alternative to residential care. However, while aging in place may bring social, psychological, and health benefits, there can also be downsides on an everyday level (Sixsmith and Sixsmith, 2008). Living at home in old age can lead to negative experiences, such as neglect, depression, isolation, and loneliness (Sixsmith and Sixsmith, 2008). This is likely to be even more important for people with dementia and mild cognitive impairment (MCI), especially those who do not have strong family and community support networks around them as they may struggle to understand and express their situation in ways which enable them to get help. This can be compounded by a lack of or inadequate culturally sensitive community support or care services, appropriate and affordable housing, and accessible family support (Fang et al., 2016). One's sense of safety and security at home and in local communities can be improved by age-friendly designs to reduce risks for accidents and injury,

sutadimages: Shutterstock

and to provide rapid responses when a person needs help. Technology-based solutions can be used to enable older adults with dementia to age well in the right place through better support for themselves, their families, carers, and service providers. Such technologies include smart home systems, assistive robots, socially connected platforms, serious games, and smart *connected* age-friendly cities and communities.

4.2   WHAT'S IN THIS CHAPTER?

This chapter considers people with dementia using a holistic perspective in terms of the technologies that can support them to age well in the right place, using personas and scenarios based on real-world examples to understand the person in their situation (Mike's persona and scenario), what needs, hopes, and desires they might have, and potential technological solutions. Supportive technologies for homes and communities are then overviewed in relation to the real-world needs of older adults with dementia using examples of work with indigenous populations (Box 4.1), home educational applications (Box 4.2), smart alert (Box 4.3), and control systems (Box 4.4). Finally, a key initiative regarding technological developments is discussed.

4.3   PERSONA AND SCENARIO—MIKE

Mostovyi Sergii Igorevich: Shutterstock

Persona: Mike is an 82-year-old widower who lives alone in a traditional neighborhood in the center of Liverpool in the United Kingdom (UK). He is generally in good health, but has been living with mild dementia for about five years, which has now progressed to a moderate level that significantly impacts his everyday activities through a general forgetfulness of recent events, losing objects, and confusion over time of day. While he is quite forgetful, he is able to live at home with the support of his sister and he can still do some things for himself. He likes to go out and get his supplies from the local shopping center. He makes his own meals and looks after the house, which is generally clean and tidy. Other than that, he doesn't do much apart from watching television and puttering about his home. Mike's sister plays an important role in his life. She visits two or three times a week to keep an eye on him and to deal with the things that Mike cannot cope with, like paying bills. Mike is a very independent-minded person who is unwilling to accept strangers and care workers into his home, preferring to look after himself with the support from his sister. He is aware of the problems he has due to dementia and is worried that he might not be able to cope and that he will be forced to leave his home and his neighborhood.

Scenario: Despite his moderate dementia, Mike does quite a lot for himself, such as shopping, and he typically goes out twice a day for an hour in and around his neighborhood. He enjoys the visits from his sister, although he often does not remember them. Over a period of a few weeks Mike started going out less and spending less time away from the home. Mike was suffering from ulcers on the soles of his feet. These ulcers made it too painful for Mike to stand, or to walk for more than short distances. Around this time Mike's sister started to visit less, because she was looking after her husband who had become seriously ill. His sister had been taking care of Mike's health needs and had actually looked after his feet by cutting his toenails. After two weeks Mike had become more or less confined to his home, unable to see anyone or even buy food, while his sister had not visited at all for a week. If this situation continues, it is highly likely that Mike's health would entirely break down. Intervention by emergency services would be likely and admission into acute hospital care. Many older adults in these circumstances are then discharged into long-term residential care because they are deemed unable to look after themselves at home. It is important to note that the major challenge to Mike continuing to live at home was not due to a change in his cognitive status, but due to a change in his support network.

Solution: One major area of work in AgeTech is home monitoring of people's activity and health. A system called Care in the Community (Sixsmith et al., 2007b) was piloted in the UK and used simple infra-red movement sensors in the home to detect people's patterns of activity, visits by other people, and when the person leaves and enters the house. This basic data was compiled to create a typical activity pattern for the person. The changes in Mike's patterns of activity would send an alert message to a call center that could then arrange community health professional to visit and arrange support services. This preventative approach avoids admission to expensive acute and residential care, while helping the person to remain independent at home.

4.4   TECHNOLOGY FOR SUPPORTIVE HOMES AND COMMUNITIES

Supporting someone like Mike to stay at home, with limited help and family care, will require new solutions in a number of key areas. Technology for supportive homes and communities has been the focus for much of the research and development in AgeTech for people with dementia, including smart homes to support independent living and age-friendly community initiatives that are increasingly putting technology at the forefront (Savoy, 2018). Much of the early research focused on home safety and security. For example, fall alert and detection systems, systems to automatically shut off devices (e.g., stoves or faucets), and activity sensors that track an older adult's movements and report them to a caregiver (Blackman et al., 2016). Technologies are also emerging that help support a person's ability to remain living at home (see Chapter 6 on autonomy and independence), such as assistive technologies. These can help people with activities of daily living around the home or medication dispensers and reminders that help older adults to manage medication independently and to ensure the can connect with important adults in their lives for social participation and advice and reassurance (for a list of promising technology solutions see Blackman et al., 2016).

Box 4.1 Technology and Dementia in Indigenous Communities

If supportive technologies are to be appropriate to the real-world needs of older adults with dementia, then research and development requires the input of those people who will use them (see Chapters 11 and 14). Research on how technology is supportive of older adults in indigenous communities is currently being conducted by the University of Saskatchewan (Saskatoon, Canada) partnered with the File Hills Qu'Appelle Tribal Council, in Fort Qu'Appelle Saskatchewan, to examine the ways indigenous communities use technology for the health care pathways in dementia care. Similarly, in Ontario, Canada, a collaborative research team of AGE-WELL NCE investigators and Anishinaabe adults are aiming to increase the accessibility and cultural safety of supportive technologies. Through knowledge mobilization and respectful dialogue these researchers are working to bring culturally appropriate supportive technology solutions for multiple indigenous communities (AGE-WELL Workpackages—Core Research Projects Listing, 2020). In addition to research projects, there are community initiatives and commercial products on the market to support healthy lifestyles.

Box 4.2 Supportive and Safe Homes Education App

The Dementia Friendly Home (DFH) App (dementia Australia™ (n.d. a)), developed in collaboration with dementia Australia, Deakin Software and the Technology Innovation Laboratory, and the joint Commonwealth and State Government Home and Community Care Program, is an educational mobile app that simulates a walk through a virtual home to better understand how to make changes to support a safe environment for a person living with dementia at home. The DFH recognizes that many individuals with dementia remain living in their homes within the community and aims to support this. The app uses 10 educational principles that recommend many ideas for small, practical, inexpensive changes that can help support an individual's regular lifestyle activities, independence, and engagement with the community for as long as possible.

Box 4.3 Silver Alert

Community ASAP is an alert system (mobile website and app that alerts and mobilizes community citizens) created by AGE-WELL NCE that helps locate people with dementia who might wander and get lost outside their homes (Daum et al., 2019). The technology is an app that allows local community volunteers to look for missing persons in their neighborhoods. This project emphasizes the need to think beyond the technology and how the technology is part of a community-based initiative. Policy development is crucial and the project team worked with a Member of the Alberta Legislative Assembly to inform an evidence-based decision on expanding Alberta's Missing Persons Legislation. The result was Bill 210: the Missing Persons (Silver Alert) Amendment Act (AGE-WELL News, 2020a), which came into force on July 1, 2018 and includes older adults with cognitive challenges. There is a petition for a national silver alert strategy before Canada's federal government to spread this policy change across Canada. A similar app/service in Scotland is called Purple Alert, and other terminologies are used elsewhere in the world.

Box 4.4 Emitto Smart-home Control

Emitto, designed by the smart tech company Novalte (2020), in Halifax, Canada, is a single device that allows people with limited mobility to control devices within their home. Emitto is a unique product in that it uses cloud-based technology to connect with household items, such as lights, televisions, automatic doors, and thermostats. For example, it allows an individual to open the front door or turn up their air conditioning without assistance from a support care worker or family. The Emitto system can be operated through a variety of accessible mediums, such as voice, accessible switches, smartphone, voice-assisted technology, and more. In this sense, Emitto can be customized to the unique needs and wants of the user, enabling older adults to be supported in their own homes on their own terms.

4.5   KEY INITIATIVE—HOGEWEYK DEMENTIA VILLAGE

The Hogeweyk Dementia Village in the Netherlands is designed to provide care for people living with severe dementia and underlying health issues. The Hogeweyk provides nursing-home care but looks and feels like an ordinary Dutch neighborhood, and its purpose is to enable people with dementia to live an ordinary life. It is a self-contained community that facilitates a safe and independent life for people with dementia. It is designed with houses, streets, squares, and gardens to offer people living with severe dementia the same freedom, experiences, and living conditions as everybody else. It shares its amenities with the local community and includes facilities such as a theatre, supermarket, restaurant, hair salon, physiotherapy annex gym, and café. Living spaces and activities are tailored to meet the interests, wants, and needs of individual residents. For example, home décor and design are based on their previous lifestyle and what they value. Residents are encouraged to manage their households together with the support workers as needed. By maintaining familiar routines, responsibilities, and lifestyles, residents are challenged and stimulated to remain active in daily life. This vision of quality of life for people with dementia is called The Hogeweyk. It has received multiple international awards, and the concepts and details showcased here have already been tailored and applied to various locations around the world (Hogeweyk Dementia Village, 2017).

Courtesy of Hogeweyk Dementia Village

4.6   FIND OUT MORE

10 Priorities Toward a Decade of Healthy Ageing. (2020)—Geneva: World Health Organization, Department of Ageing and Life Course: https://www.who.int/ageing/10-priorities/en/. Is a brochure that identifies 10 action priorities for attaining objectives of global strategies and presents an action plan on aging and health developed by the WHO.

Active ageing: a policy framework in response to the longevity revolution—This report, edited by Paul Faber, presents a policy framework based on active ageing. Rio de Janeiro: International Longevity Centre Brazil: http://ilcbrazil.org/portugues/wp-content/uploads/sites/4/2015/12/Active-Ageing-A-Policy-Framework-ILC-Brazil_web.pdf.

Thinking about aging in place—This brochure provides guidance for older adults to succeed in aging in place, developed by the Government of Canada. Gatineau: Human Resources and Skills Development Canada: https://www.canada.ca/content/dam/esdc-edsc/documents/corporate/seniors/forum/place.pdf.

CHAPTER 5

Health Care and Health Service Delivery

5.1   THE CHALLENGE

While most older adults enjoy active and healthy lives, declining cognitive health is a reality for many as they grow older. Providing the best possible health care for older adults with dementia is a major policy priority in high income countries (HICs), and is becoming more critical as population aging is experienced in low- to middle-income countries. Much of the care provided to people with dementia who can no longer manage at home is in long-term residential settings. While these provide protective environments for the most frail, some argue that institutional care is typically just warehousing by providing the basics of safety, security, and supporting the tasks of daily living (Kroemeke and Gruszczynska, 2016). There is an urgent need for training and care that is more holistic and person-centered, i.e., in that it involves people with dementia as partners in the planning, development, and assessment of their care needs so that care is provided in the most appropriate way to support better quality of life for residents (Kim and Park, 2017). In the community, services have often been under-resourced and under-developed and it is important to

Melinda Nagy: Shutterstock

provide more preventative and community-based alternatives to long-term residential care that help older adults manage their own health conditions and live independently at home. One of the major challenges facing countries across the world is how to create sustainable services in the face of increasing numbers of older adults and the associated greater health and social care financial and human resource cost. In many ways, this challenge is more about making better use of the financial resources and human capital that are already in place, ensuring equitable access, and delivering care that is responsive to individual needs and preferences. Technology-based solutions can be used to improve implementation of services and service delivery as well as health assessment, monitoring and treatment, including telemedicine and telecare systems, medication monitoring and reminding, and wearable devices to track health status.

5.2   WHAT'S IN THIS CHAPTER?

These issues are exemplified in the persona of Dorothy and her situational context. Next, a possible technological solution is introduced and the potential role of technological supports in the health service are discussed. Several examples of such supports are presented: Pill dispenser (Box 5.1); Home4Care (Box 5.2); Enabling Educational Dementia Immersive Experience (Box 5.3); and information about further research initiatives are described on data-driven health care and technology for supporting cognitive health assessment.

5.3   PERSONA AND SCENARIO—DOROTHY

nafterphoto: Shutterstock

Persona: Dorothy is an 89-year-old woman who has recently moved to an assisted living facility in Edmonton, Canada. She is generally in good health, but has now been diagnosed with moderate dementia, which she has been living with for about four years. Her condition and abilities have rapidly worsened over the past year and she can no longer safely live in her own home and has moved into a long-term residential care home. The activities and hobbies Dorothy used to enjoy, such as playing her piano, reading novels, and tending to her garden are now largely inaccessible to her, leading to feelings of loss, anxiety, and depression. Dorothy is considered physically healthy, but recently her mental health has declined. The facility care-providers have tried to involve her in social events and involve her in the activities conducted in the residential home, but Dorothy has largely resisted these measures and choses to sit in her room. Her family has tried to alleviate her feelings of depression by visiting her more often, but this has not improved her mood. Dorothy's family fears her decline in mental health will impact other aspects of her physical health such as her appetite and physical activity. They worry that these changes may already be happening as Dorothy has recently started losing a lot of weight. Dorothy's dementia is quite advanced, but she can still do some things for herself. She enjoys picking out the clothes she would like to wear for the day and likes to walk through the communal garden when the weather is nice.

Scenario: Dorothy has become increasingly socially isolated. Although she sometimes eats her meals in the dining room with the rest of the residents, she often choses to eat alone with minimal social interaction since her verbal communication skills have declined. Dorothy's immediate and extended family—sons, daughter, grandchildren, and friends—visit frequently, but express frustration among themselves that it is difficult to communicate meaningfully with Dorothy.

Solution: Dorothy's daughter, Michelle, has recently bought a tablet device for her mother after she read about a mobile app that may assist in organizing visits and information between caregivers. Based on the features of the app such as nonverbal picture response buttons, daily care tips, suggested activities for stimulating interaction, and a visual psychological symptom tracker, Michelle hopes that this app will lead to more informed and engaging visits for her mother. Within three months, Dorothy's caregiving team have been able to use the app in a variety of ways. While visiting, Michelle had struggled to engage her mother in conversation. Instead of feeling frustrated, using the display of various images of activities had helped improved communication and Michelle felt more in touch with her mother. For example, seeing the image of the garden had made Dorothy visibly happy. From this interaction Michelle was able to suggest a walk in the garden with her mother. Similarly, when reviewing Dorothy's psychological symptoms over several weeks, a friend noticed that Dorothy's angry outbursts when going to bed were linked to specific new bedsheets. After changing back to the original style of bedsheets the outbursts stopped. Importantly to Michelle, the new app has allowed for better health tracking on a day-to-day basis, providing a reliable tool to store information that can be used to interpret Dorothy's needs, wants, and behaviors. It can, additionally, inform professional health care providers if any care-related changes arise.

5.4   TECHNOLOGY FOR HEALTH CARE AND HEALTH SERVICE DELIVERY

Research in the area of health care and service delivery is wide-ranging. It encompasses medical records systems, systems for managing chronic diseases at home, emergency and alert monitoring systems, medication management, smart environments to manage transitions in care, to name a few. This area has been a focus for AgeTech research for many years and while it is not possible to provide a systematic summary of the area herein, it is useful to highlight some examples of the diversity and potential of technological supports in this sector.

Data-driven health care: Advancing mobile technology and off-the-shelf smart devices provide a unique opportunity for data collection for a variety of users, especially the older population, on a variety of health variables such as: heart rate, sleep habits, and activity patterns. Other vitality measures may prove useful in informing health related decisions, as well as overall health monitoring and treatment. One research group, based in the University of Waterloo, Waterloo, Canada, is aiming to develop and prove the feasibility of home health monitoring and data-driven decision support systems based on various data sources, including health records and wearable sensors. The outcomes of their projects include: (1) developing a knowledge base of mobile Health (mHealth), machine learning, and AI to be applied to healthy aging; (2) using wearable data to create frailty prediction algorithms (Yang et al., 2019); and (3) creating a health care decision support algorithm to be utilized by interRAI—a global consortium for health policy (Lee and Hirdes, 2020).

Technology for Supporting Cognitive Health Assessment: The collection of big data and the ability of machine learning and AI to detect patterns within these data sets paves the way for new solutions for identifying health care needs in a variety of populations; particularly, but not exclusively, research projects that are focusing on supporting cognitive health are harnessing these big data prospects. Some examples include a Canadian initiative at the University of Regina, Regina, Canada, led by Dr. Thomas Hadjistavropoulos (see Figure 5.1). The project is developing innovative and affordable technological solutions that can be implemented in long-term care facilities for the improvement of pain management. Specifically, by using facial recognition technology, Dr. Hadjistavropoulos aims to better to detect pain in people with dementia, which is hugely important for those who may not be able to communicate about their pain effectively (Hadjistavropoulos et al., 2018).

Implementing reliable, valid, and easily repeatable cognitive assessments is important for accurately diagnosing cognitive status and impairments. However, self-report or informant reports each have their own drawbacks and in-clinic simulations of everyday cognitive activities can be both time consuming and costly. A collaborative research project from the Wake Forest School of Medicine, Winston-Salem, U.S., aims to address these issues with the innovative approach of using computer-based simulations of everyday activities (Ip et al., 2017). The project entitled SIMBAC (SIMulation-Based Assessment of Cognition) is developing computer tablet-based simulations of

Figure 5.1: Highly qualified personnel (HQP) at Dr. Hadjistavropoulos' lab have coded manually over 1,000,000 video frames of older adults suffering from pain in order to inform and develop automated pain detection algorithms. Image courtesy of Thomas Hadjistavropoulos.

five common yet cognitively demanding everyday activities, ranging from remembering faces to re-filling prescriptions via telephone. These simulations provide a repeatable, accurate, and timely tool for assessing cognitive abilities. Further, SIMBAC provides preliminary evidence that results from simulated testing correlated with proxy-reported functional status, and correctly classifies levels of cognitive impairment (Not Impaired, Mild Cognitive Impairment, and Dementia).

A number of initiatives and products are available to support health and health service deliv-ery, with an increasing focus on community-based services are described in Boxes 5.1, 5.2, and 5.3.

BOX 5.1 Pill Dispenser

Electronic pill dispensers and reminding systems have been developed by a variety of providers with an assort-ment of functional utilities, depending on an individual's needs. Most commonly these dispensers are equipped with a programming option to alert the user at programmed times to take a specific loaded compartment of medication. Alarms can be as simple as flashing lights and sounds to more complex systems that push notifica-tions to smartphones or watches. Notifications and summary reports can also be sent to caregivers. While these pill dispensers can help to ensure that the medication will be removed from the box, thereby improving auton-omy of the individual, there is currently no assurance that the medications will actually be ingested.

BOX 5.2 Home4Care

Funded by the National Institute on Aging, in the United States, Home4Care is a mobile application designed for improving quality of life at home for persons with dementia (Hearthstone, 2017). While still in the final stages of development, the Home4Care app is intended to include four different components to ensure high quality care for persons with varying severity of memory impairment. These components will include meaningful activities carers can use to interact with their clients or family members, dementia training for caregivers, personal preferences page, and task planning and tracking. An overarching goal of the app is to provide caregivers with a user friendly and accessible resource about their client or loved one that will facilitate higher-quality, person-centered care delivery.

BOX 5.3 Enabling EDIE

Developed by dementia Australia (n.d. b), Enabling EDIE (Educational Dementia Immersive Experience) is a unique virtual reality workshop that immerses the participant in a virtual environment that enables them to see the world through the eyes of a person living with dementia. Designed to give caregivers an enhanced perspective of what it may be like to live with dementia, EDIE helps caregivers think critically about identifying support needs in partnership with a person living with dementia. The program is intended to enhance the participants' knowledge about living with dementia while exploring how to develop an effective dementia support plan that focuses on enablement.

5.5   KEY INITIATIVE

REACH (Responsive Engagement of the Elderly Promoting Activity and Customized Healthcare) is a large European Union funded project developing a service system to turn clinical and care environments into personalized modular sensing, prevention, and intervention environments/systems that support older adults healthy via activity (physical, cognitive, mobility, personalized food, etc.). The consortium has 17 partners from higher education institutions, along with the industry partners (including leading European health care technology, rehabilitation, and care and hospital firms). Figure 5.2 illustrates the kinds of health-enhancing technologies developed by REACH.

Figure 5.2: ActivLife system ready for use at care homes and rehab clinics (photo: Alreh Medical).

5.6   FIND OUT MORE

World Health Organization—Health systems and service delivery. Is a website that describes health systems delivery and offers links to how they are being implemented, and advocates for changes globally.

World Health Organization (2020)—Health systems service delivery: http://www.who.int/healthsystems/topics/delivery/en/.

World Health Organization—Ageing and health: Key facts (February 5, 2018). Is a website that offers important facts about aging and health: https://www.who.int/news-room/fact-sheets/detail/ageing-and-health.

Government of Canada/Canadian Institutes of Health Research, Personalized Health/Personalized Medicine—A website linking to current initiatives of personalized health care and e-health strategies: http://www.cihr-irsc.gc.ca/e/50119.html.

CHAPTER   6

Autonomy and Independence

6.1   THE CHALLENGE

Governments worldwide have developed policies aimed at supporting older adults to stay living independently at home and in the community for as long as possible (Garvelink et al., 2017). In this respect, independence, as a multidimensional concept, refers to functional ability (regarding both activities of daily living and instrumental activities of daily living) in everyday life, remaining active and contributing to the community (Dwyer and Gray, 2000), with financial security and social and psychological resilience. This relates to the notion of maintaining autonomy, defined as "the right of the individual to be self-determining and to make independent decisions about his or her life" (Kanny and Slater, 2008, p. 194). However, age-related physical and cognitive changes, and dementia, can undermine a person's ability to live independently and carry out everyday activities and tasks. Assistive devices and supports to help older adults, especially those experiencing cognitive decline and dementia to live independently can have major benefits for well-being and self-esteem and often reduces demand on caring services. Technology-based solutions include fall detection and alert systems, assistive technologies to help with everyday tasks and rehabilitation technologies. It is important to remember that even people with very severe dementia can enjoy doing things that will give them pleasure. Thus, technology developers should not just focus on the basic instrumental tasks of everyday life.

6.2   WHAT'S IN THIS CHAPTER?

This chapter uses the persona and scenario of Amrita to identify the complex issues that need to be taken into account to enable independent living at home in the community for as long as possible. The development of technological solutions to support independent living are discussed with specific examples of a hydration reminder (Box 6.1) and digital clocks (Box 6.2). Finally, the chapter presents the INDEPENDENT project as a key initiative in the UK that developed technologies to help older adults with dementia engage in enjoyable activities in their everyday lives.

6.3   PERSONA AND SCENARIO—AMRITA

Persona: Amrita is a 79-year-old widow, who lives with her extended family in a large house in the Vancouver area of British Columbia, Canada. Amrita is originally from India and emigrated with

Teja Sv: Shutterstock

her family to Canada over 25 years ago. Family life, local South Asian culture, and community have always been important to Amrita. Recently, Amrita has noticed that she has become increasingly forgetful although this is not a major problem for her as she has strong family support. Since the death of her husband 15 years ago, Amrita has taken on more responsibilities within the household, such as managing the finances and maintenance of the house. Her family offers to help where they can to alleviate some of the physical and mental stress, telling her that she needs to take time for herself, but she has often been heard saying "I'll rest when there's no work to do," which, according to her, will never happen. Because of Amrita's busy everyday life and priority focus on her family rather than herself, Amrita has neglected to visit her family physician for many years. As Amrita has been becoming more forgetful in recent years her daughters have tried to convince her to seek medical advice. Amrita declines each time it is suggested, insisting that is just old age. However, she has recently confided in a close friend that she is scared of what may happen to her ability to stay independent at home if she goes to a doctor to address the concerns. Amrita loves to cook and is the main cook for her family and also makes food for the local community temple, where she prepares traditional South Asian food for local older adults. Cooking is something that gives her a lot of satisfaction.

Scenario: Even though Amrita is getting on in years, she is able to make an important contribution to family life and her local community through her food. Over the last year, she has found cooking more and more challenging saying: "I have always been a great cook, but now everything goes so slowly." This makes her feel quite irritated because she has always been used to having her hands full. She feels frustrated because things that she could do easily before are now taking her a long time. It took her a whole day to prepare a dish that previously would have taken an hour because she had difficulties remembering where she was in the process. Her family and friends try to help, but sometimes she gets angry and shouts at them, which later make her feel upset and guilty.

Solution: After having an argument with her daughter over the preparation of dinner, Amrita agrees to try to find a memory aid to help with her cooking. Amrita has never used written recipes to create her meals, but she now uses an app on her tablet that enables her to write the steps of her own recipes. The app provides a visual interface that checks off steps as the recipe progresses and can read instructions out loud with the use of voice prompts. Amrita enjoys using the app and is pleased that she is able to cook more quickly and independently again. This solution helps Amrita to feel that she is "still in control" over things and is not reliant on others for help.

6.4   TECHNOLOGY FOR AUTONOMY AND INDEPENDENCE

The development of assistive devices to support activities of daily living has been an important area of research and development in the AgeTech sector. Much of the early research in this area focused on people with physical and mobility impairments, along with people who had sensory impairments. More recently, attention has begun to focus on supporting the autonomy of people who are experiencing cognitive decline or dementia such as:

1. smart technologies that aids instrumental activities for daily living (IADLs) for older adults living in the community (Rogers and Mitzner, 2017);

2. telepresence technologies that aid in connecting older adults with family and friends making it seem like the person you are communicating with is in the room with you (Rogers and Mitzner, 2017);

3. technologies, such as apps, that help prompt or remind older adults with memory problems about medications and appointments;

4. technologies that enable medical care through telepresence systems (Rogers and Mitzner, 2017); and

5. technologies to support mobility and orientation for people with dementia in the environment using GPS systems while addressing issues of autonomy and independence (Wan et al., 2016).

Innovative research, spearheaded by the Université de Sherbrooke and the University of Toronto, and partnered with CrossWing Inc., a research group in Aurora, Ontario, and Vigilent Telesystems Inc., in Dorval, Quebec, are working toward developing a mobile robot to encourage and help people perform activities of daily living. This robotic system, coined VIGIL, would allow for virtual visits between individuals and their health care providers (Michaud and Nejat, 2020). In addition, VIGIL will be able to assist older adults with their everyday needs, providing advice on basic household tasks such as meal preparation, exercise/therapy, self-care, and scheduling. Practical for many older adults, this research could be highly beneficial in maintaining independence for older adults who may be experiencing cognitive difficulties (Michaud and Nejat, 2020).

The ability to successfully complete activities of daily living is an important facet of being able to live independently. Older adults who experience progressive declines in cognitive function often struggle with completing activities of daily living. Researchers at the University of Toronto aim to develop an assistive system that will specifically facilitate the activities of daily living of dressing. This is important because dressing is often an activity of daily living that is affected by dementia and an inability to get dressed can prevent people getting up in the morning and progressing with their day. It is proposed that the development of a socially assistive robot will enable the selection of appropriate clothing for various weather conditions, activities, and personal preferences. Through continuous monitoring and perception systems, this robot is intended to provide feedback to indicate dressing failures and coach people to correct dressing failures. The smart clothing robot is expected to improve quality of life for older adults by promoting independence and autonomy.

Persons with dementia and mild cognitive impairment can often experience difficulties in remembering to complete everyday tasks or carry out specific responsibilities. The emergence of the Internet of Things (IoT) and connectivity of devices (e.g., Bluetooth®, Radio Frequency Identification (RFID), and Near Field Communication) offers the opportunity of a more integrated smart environment to support people in their everyday lives. Jawaid and McCrindle (2015) have developed a framework that uses smartphones and sensor technologies to integrate data, aptly named computerized help information and interaction project (CHIIP). CHIIP will facilitate the aggregation and delivery of information from supportive technologies to help individuals to perform relevant tasks and activities in their home or local environment.

Supporting the autonomy of people with dementia using assistive devices has attracted much commercial interest and consequently there are an increasing number of devices available in the market, available through online retailers and mainstream electronics stores and pharmacies. Two such examples are highlighted in Boxes 6.1 and 6.2.

BOX 6.1 Hydration Reminder

Maintaining adequate levels of hydration is problematic for individuals with severe cognitive impairment. Droplet® is designed specifically to assist older adults with dementia by ensuring they can independently maintain good hydration levels. Using an electronic smart base paired with a Droplet® tumbler or mug, friendly customized voice recordings and lights are used to remind individuals to consume water. Droplet® can be customized for the needs of the individual by including handles on the mug, a flow control lid, or remaining as a traditional glass. Droplet® also features a light-detecting function to enable a nightlight feature to assist individuals to find their drink in the dark, such as on a bedside table during bedtime.

Figure 6.1: Example Droplet® tumbler configuration (from website).

BOX 6.2 Digital Display Clocks

Digital clocks are simple useful devices for supporting time orientation for persons with dementia, helping them to conduct daily activities at appropriate times and are readily available in stores (The Alzheimer's Store, 2020). These digital clocks can range in their functionality and display settings, with the simplest displaying only the time and the date. More complex devices can include changing backgrounds that reflect the time of day it currently is, displaying a bright shining sun for day time hours then gradually changing to a starry nighttime sky. Additional features can include voice reminders for specific activities, connections to smartphones to send notifications, photos and other messages, and interfacing with other smart home devices. These digital clocks, in their varying complexity, can help individuals maintain reality orientation in order to minimize disruptions in their daily activities.

6.5   KEY INITIATIVE—THE INDEPENDENT PROJECT

The INDEPENDENT project was funded by the UK's Engineering and Physical Sciences Research Council (EPSRC) to develop technologies to enhance the quality of life of people with dementia. The INDEPENDENT project worked to involve people with dementia in all stages of the project from early product ideas, through prototype development, and then field trials. This highlighted the importance of everyday activities such as listening to music, social activities, and having conversations, but also uncovered major barriers (Sixsmith et al., 2007a). For example, many people enjoyed listening to music and it was an important source of social connection, as well as activities within and outside the house, but found that music players were typically too complicated to use (see Figure 6.2).

kuzmaphoto: Shutterstock

INDEPENDENT went on to develop a number of prototype technologies (Orpwood et al., 2007), including a Simple Music Player to allow people with dementia to easily access and listen to the music they like (Sixsmith, Orpwood, and Torrington, 2010). The device was successfully trialed in residential homes and in the homes of people who lived in the community (Sixsmith and Gibson, 2007).

Figure 6.2: Simple Music Player prototypes.

A major outcome of the INDEPENDENT project was a commercial Simple Music Player. The prototype Simple Music Player was commercialized by designability, a not-for-profit organization in the UK that aims to create and market attractive and usable assistive technologies (designability, 2020). Designability was the key to taking the prototype device to market, both in terms of final product development and the marketing of the device through the E2L Consultancy Group (2015). Many of the features of the original prototypes have been included in the commercial product, including a simple on-off interface and prompt to encourage people to listen to music.

Sales of the Simple Music Player have now exceeded several thousand units and it is available via distributors worldwide, including in the U.S., Scandinavia, and Canada. The Simple Music Player is part of the product line at major pharmacies in London (UK). Many of the ideas from INDEPENDENT were incorporated in further research, notably the development of a Musical

Chair for use in residential homes and the European Union funded the SOPRANO project that aimed to develop ambient assistive living technologies for older adults.

Figure 6.3: Simple Music Player provides an easy way for people living with dementia to enjoy their favorite music. Courtesy: Designability.

The success of the Simple Music player shows the value of involving people who are "hard to reach" or "seldom heard" in all stages of the research and development of a technology, despite the significant challenges the research led to real-world products that now benefit the lives of people with dementia. Researchers need to *listen* carefully to understand the true desires and preferences of people and uncover how people with very severe cognitive impairments can still enjoy the simple things in life—like music!

6.6   FIND OUT MORE

The Alzheimer's Society of Canada provides guidelines to help with buying assistive devices— https://alzheimer.ca/en/Home/We-can-help/Resources/Shopping-for-assistive-products.

Barrett, L. (2014). Home and Community Preferences of the 45+ Population 2014—Is a report discussing the home and community preferences of the 45+ population and investigates issues for middle aged and older adults regarding their home and community life: https://www.aarp.org/content/dam/aarp/research/surveys_statistics/il/2015/home-community-preferences. DOI: 10.26419/res.00105.001.

Further information on the Simple Music Player is at—http://www.dementiamusic.co.uk.

Lothian, K. and Philp, I. (2001). Maintaining the dignity and autonomy of older people in the healthcare setting—*BMJ* (Clinical research ed.), 322(7287): 668–670. DOI: 10.1136/bmj.322.7287.668.

Rogers, W. A. and Mitzner, T. L. (2017). Envisioning the future for older adults: autonomy, health, well-being, and social connectedness with technology support—*Futures*, 87: 133–139. DOI: 10.1016/j.futures/2016.07.002.

CHAPTER 7

Mobility and Transportation

7.1 THE CHALLENGE

Mobility is essential for older adults' continued health, social participation, and quality of life. Individuals require convenient and appropriate options to meet their mobility and transport needs, both within and outside their homes and communities. This need for accessible mobility options is just as important for persons with dementia, as mobility in and around the house is the key to independent living for all older adults and aging in place. However, for people with dementia, getting lost outside the home can be a major problem and technologies to restrict their mobility to promote safety have become a significant development in recent years. As was discussed with the Silver Alert system (Chapter 4), technologies to help find people when they get lost are increasingly used. Evidently, there is a continued need to develop resources that balance the facilitation of autonomous mobility and safety for people with dementia.

Helping people with varying severity of cognitive decline, including dementia, to get around needs to be a priority. As Woolrych et al. (2019) have shown, mobility is integral to older adults' social participation in everyday life, however good transportation links in themselves are not enough to reap the psychosocial benefits of social participation. Being able to navigate around local streets, communities, as well as both town and city centers has been linked to active aging. Affordable, accessible, and usable public transportation is also seen as an important aspect of age-friendly cities (WHO, 2007). Appropriate street furniture (e.g., benches, information on bus timetables) and the availability of public toilets are essential environmental and informational aspects of age-friendly design supporting mobility and the inclusion of older adults, including people with dementia.

For many older adults, particularly in high income countries (HIC), driving is the main means of transportation. The ability to drive is linked to maintaining independence as well as to accessing health, social care, and

lightpoet: Shutterstock

leisure services. While most older adults continue to drive safely, many eventually have to stop due to declining physical, mental, or sensory abilities or for financial reasons. In terms of cognitive impairment or dementia, confusions in wayfinding can constitute a major obstacle to mobility and the use of various modes of transportation.

Losing mobility, whether through walking, using public transport, or driving, can severely constrain an older adults' lifestyle and negatively impact their mental and physical health. This is more likely to occur as the person becomes frail or develop physical and cognitive disabilities, such as dementia. At this point, public services and family are often required to support the older adults' needs, and this can be both burdensome and costly. For example, according to the American Association of Retired People (AARP), 78% of caregivers provide or arrange for the transportation needs of older adults every year (Feinberg et al., 2011; NAC and AARP, 2015; The National Alliance for Caregiving and AARP Public Policy Institute, 2015), with a recent study estimating that caregivers provide 1.4 billion trips a year (AARP, 2011). Development of technologies to support the mobility and transportation needs of older adults, including people with dementia, can have many benefits: improvements in fitness and physical activity, enabling access to health and social care services, leisure opportunities, contributions to family life, and generally reduction in loneliness and social isolation. Technology-based solutions to mobility and transportation needs include smart mobility devices and wheelchairs, way-finding apps and, in the future, autonomous vehicles.

7.2   WHAT'S IN THIS CHAPTER?

This chapter introduces the complex intertwining of the daily routines, functional needs and social participation of older adults' lives, and potential technological solutions when everyday life becomes more difficult to navigate, through the persona of 70-year-old Eva. Research and development of technologies to support mobility, and access to and use of transportation, when cognitive decline makes home-based tasks (aids to support home based tasks), travel (Box 7.1, smart wheelchairs, mobility scooters, smart vehicles), and wayfinding (wearable navigation devices) difficult are then discussed. Such technologies can help people to manage their everyday life as well as enjoy social and cultural life (Box 7.2, travel and tourism). Since the positive impact of such technologies on older adults' health and well-being is implicated in such technologies, more research is needed to establish the health benefits of such technologies.

7.3   PERSONA AND SCENARIO—EVA

Persona: Eva is a 70-year-old woman who lives alone in a small village in the south of Sweden. Her only daughter lives in Stockholm with her family, but Eva likes to visit them every few months and travels by train to stay with them. She likes to see her two little grandchildren, but cannot stay too long because her daughter's apartment is small and she does not wish to be a nuisance. Eva has no personal complaints about her health and overall thinks she has good health habits. Eva's doctor recommends that she should be eating fewer sugary snacks and junk food, but Eva can't help but pick up a treat for herself when she goes out on her day trips. Eva likes to travel around to get her shopping or just to get out of the house. She uses public transportation, which is excellent and connects her with other villages and towns nearby. Sometimes she catches the train for a day out in the nearby cities of Malmö and Copenhagen.

Ruslan Huzau: Shutterstock

Scenario: One day Eva decides to visit a small town near her village, a place that she visits frequently for shopping and to have a cup of coffee and a confection in one of the cafés. She catches a bus outside her house and gets off at the usual stop in town. Suddenly Eva does not know where she is. She walks around but does not recognize any of the places she sees. Eva feels awful because she cannot understand what is happening. She begins to panic. Eva does not remember much about what happened after that. She still does not know how she got home. She thinks she went into some stores and asked people where she was and she presumes someone helped her to catch a bus back home. Eva is frightened by the whole experience and is afraid that it will happen again when she goes out. This is a big problem for Eva as she needs to get the bus into town to buy her groceries or visit her daughter.

Brian A. Jackson: Shutterstock

Solution: Digital walking maps can help people find their ways to and from places, as well as using public transportation. Such maps should include pictures to help people recognize familiar places, while voice interfaces can help to reassure people and prompt them to find their way. It is important even to include simple information such as a home address in case the person forgets.

7.4   TECHNOLOGY FOR MOBILITY AND TRANSPORTATION

Technology offers huge potential to enhance the mobility in the home and community of older adults who are experiencing cognitive decline, including older adult friendly staircases, smart homes, mobility scooters, public transportation information systems and technologies that maintain older adults' ability to operate a car, or car sharing services and automated vehicles. Improving mobility hugely contributes to many aspects of a persons' life including access to health and social services, leisure events, and work opportunities, enhancing both autonomy and independence. Research at the University of British Columbia, University of Waterloo and Université de Sherbrooke is focusing on the development of intelligent scooters and wheelchairs to maintain autonomy through mobility. The research team aims to design for older adults who are experiencing physical, perceptual or cognitive ability declines or limitations, which may make it difficult to learn the driving mechanics of a traditional motorized assistive device. CoPILOT (Collaborative Power Mobility for an Aging Population) is a proposed solution through the development of intelligent control technologies that will compensate for individual's limitations operating a motorized scooter or wheelchair to promote independent mobility (Miller and Mitchell, 2020).

Older adults can experience mobility difficulties in performing everyday tasks such as carrying weighted items up and down stairs. The Quantum Robotic Systems (QRS) research team is working on a project specifically designed to aid older adults with this issue. QRS's Robotic Stair-Climbing Assistant ("ROSA") is an assistive device meant to aid older adults in a residential setting. ROSA is designed to be an easy to use robotic cart that can autonomously carry household items (laundry, boxes—up to 100 lb) up and downstairs. Further considerations of how automated systems like this could support the daily activities of people with dementia should be investigated.

The use of wearable devices to support outdoor mobility for older adult users is a growing field of research. Investigation into what these devices might look like, the data they collect, and how they operate to support older adults in various stages of cognitive decline are just some of the contextual scenarios researchers are exploring. In Europe, a group of researchers have identified key features and considerations for assistive technology devices for outdoor mobility. They find that the technology should: (1) automatically adapt to environmental extrinsic factors for safety; (2) adapt to intrinsic factors, such as psychological stress; and (3) account for long-term intrinsic patterns, such as favorite mode of transportation, social routines, and scheduled activities (Teipel et al., 2016).

The development of smart vehicle technologies such as parking assistance, crash avoidance, collision detection, lane-assist, built-in GPS, and other automated features are on the rise for the affordability of and inclusion in new vehicles. These developments create a unique opportunity to support older adults' mobility and autonomy of transportation needs. Many older adults who experience cognitive decline face eventual driving cessation. Understanding how new driving technologies could be implemented and used to extend independent driving and mobility is important.

Recent research groups are exploring older adults' opinions concerning these technologies (Freed, Ross, and Stavrinos, 2019; Oxley et al., 2019), as well as groups that are further developing the technologies with cognitive health changes in mind (Knoefel et al., 2019).

BOX 7.1 Smart Wheelchair

Figure 7.1: Dr. Pooja Viswanathan, founder and CEO of Braze Mobility, with the company's navigation solution for wheelchair users. Courtesy: Braze Mobility.

Braze Mobility is a Canadian company working in the mobility and transportation challenge area (https://braze-mobility.com/). Braze Mobility is a start-up supported by AGE-WELL NCE that has created an add-on system that will turn any powered wheelchair into a smart wheelchair to help avoid collisions and enhance navigation, which is particularly useful for people with dementia (AGE-WELL News, 2020b). The sensors attached to the wheelchair automatically detect various types of obstacles and provides multisensory alerts, including audio, visual, and vibration, to avoid incidents. Braze Mobility's versatility in sensor placement and feedback allow for many different mobility wheelchair users to benefit.

BOX 7.2 Travel and Tourism

Travel and tourism contribute substantially to the global economy and research indicates that persons with dementia have the same desire for travel as others. Researchers at the University of South Wales, Australia have collected empirical evidence of how assistive technology could be used to support persons with dementia in their travel and tourism endeavors (Asghar et al., 2019). Their research has found that independent travel is desired by people with dementia and that the accessibility of assistive technology can aid them in doing so.

7.5   KEY INITIATIVE—UNIVERSITY OF SOUTHERN CALIFORNIA

The University of Southern California (USC) is using their recent results from a study on ride-sharing usage for older adults to inform and advocate for policy changes to support older adults' mobility through the use of ride-sharing. Supported through a grant by the American Association of Retired Persons (AARP) and partnered with a ride-sharing service Lyft (Lyft Inc., San Francisco, U.S.), the USC aimed to investigate the impact of networked transportation on overall health for the older adults participating in the study. In order to tackle the critical health concerns that can arise when older adults are isolated by their mobility issues. This USC study investigated how breaking down the barriers of technology use for network transportation impacted multiple aspects of health. The study implemented three months of free and unlimited networked transportation though the ride-sharing company Lyft. Additionally, the study provided the participants with personalized training on how to use the mobile app and the call-in service for accessing rides.

While some older adults were initially wary of the app and utilized the call-in service, many later migrated to the app because it was more convenient, in terms of utilization and time. Lyft's vice president of health care (note: it is innovative that they even have a vice president of health care) stated: "This validates what many studies have found: that older adults are motivated to break down barriers to improve their health, and once educated, technology is no longer a barrier" (Snelling, 2019). The results of the study are promising for what increasing access to mobility and transportation can do for the overall health of older adults. As a result of the increased access to transportation participants reported easier access to health care, ability to fulfill medical appointments, increased social engagement, and improved subjective quality-of-daily-life (Saxon, Ebert, and Sobhani, 2019). Additionally, participants who reported that they had stronger social support networks took fewer rides, suggesting those who are socially supported may have required fewer due to their transportation needs being fulfilled by their own network (Saxon, Ebert, and Sobhani, 2019).

The conclusion of this pilot study found that majority of older adults would like to continue using the network transportation but that cost would be a large barrier for them. Identifying this issue has directed the future directions of this initiative. Both the study's researchers and the president of the AARP Foundation believe that the study has the potential to broaden collaboration through private companies to develop economic models and create affordable solutions to providing network transportation for older adults. For example, one solution could be changes to insurance coverage to allow for non-emergency transportation to be covered. The nature and results of this study open the door to further conversations concerning how existing mobility technology can be used to support the health of older adults, including those who may already be experiencing cognitive decline or dementia.

7.6   FIND OUT MORE

Coresight Research (June 21 2016)—The Silver Series III: technology for mobility-constrained seniors.

Managing mobility—Transportation in an aging society: this brief discusses older adults' transportation needs and the inadequacy of current transport services and infrastructure to meet those needs. How demographic differences affect mobility. Briefing by Daniel Munro of the Conference Board of Canada (2016): https://www.conferenceboard.ca/e-library/abstract.aspx?did=8293&AspxAutoDetectCookieSupport=1.

Canadian Institutes of Health Research (March 16, 2007)—Mobility in Aging Initiative: priorities for research and research-advancing activities identified through consultations: http://www.cihr-irsc.gc.ca/e/33610.html.

CHAPTER 8

Healthy Lifestyles

8.1 THE CHALLENGE

In Chapter 2, we discussed the idea of cognitive health as the ability to perform the range of mental processes that are necessary for everyday life and some of the risks to good cognitive health in later life, including genetic factors and lifestyle factors, such as smoking and social disconnection. There is increasing evidence that a healthy lifestyle in general contributes to good cognitive health (Centers for Disease Control and Prevention, 2009). In this chapter, we discuss healthy lifestyles in terms of positive and conscious lifestyle choices that promote and sustain physical and cognitive health, including nutrition, substance use (e.g., smoking, alcohol), physical activity, and self-management of health issues. More generally, social engagement and being active are important for developing and maintaining optimal health across the lifespan. Having a healthy lifestyle is often seen in terms of individual actions, but there are important social, economic, and contextual factors that contribute to health outcomes including social isolation, physical environment, socioeconomic status, education, access to information, and culture. Technology-based solutions include virtual exercise systems, wearable fitness tracking, health and fitness apps, and information and decision tools that support a healthy lifestyle. Promoting and encouraging a healthy and active lifestyle is an essential component of a preventative agenda for health care, helping people to continue living at home (where most people prefer) and avoid admission into expensive acute-care or long-term residential care.

8.2 WHAT'S IN THIS CHAPTER?

This chapter presents dilemmas concerning the development and maintenance of healthy lifestyles as preventative interventions. This is particularly difficult when living with dementia, as the Mary-Anne and Betty persona and scenario indicate. Technological solutions are presented and discussed in relation to their fit in everyday lives and social and financial contexts.

8.3 PERSONA AND SCENARIO—MARY-ANNE AND BETTY

Persona: Mary-Anne is a 65-year-old woman who lives with her long-time partner Betty in a large city in the northeast of the United States. Mary-Anne has had a very varied life and career, having worked as a mail-delivery person, a clerk at a floral boutique for a number of years, and now working part-time as a shuttle driver for the local car dealership. She has always been an on-the-go type of person and enjoys socializing with people. Betty has worked all her life as well, but in recent years her cognitive abilities have declined significantly. She was diagnosed with early-stage dementia and is no longer able to work and does not like to go out on her own and she has grown extremely dependent on Mary-Anne for most things. Mary-Anne feels stress and anxiety under the pressure to make ends meet financially for her and Betty. She takes medication to control her high blood-pressure. She is worried about her own health and also worries what would happen to Betty if she was unable to care for her and support her financially. Mary-Anne and Betty have a quiet life together and spend most evenings at home watching TV. On occasion they will go out for a special dinner or event but with limited disposable income they prefer to find entertainment at home.

Monkey Business Images: Shutterstock

Scenario: Mary-Anne and Betty have not had the opportunity to put aside substantial funds for their retirements, but still have enough to get by, having just enough income to pay their bills and put food on the table. With Betty's dementia, their reduced household income and medical bills, the caring and financial strain has fallen on Mary-Anne who has picked up extra hours driving to help balance the costs. Unfortunately, this means that Mary-Anne has even less time for social- or health-related activities. Mary-Anne used to work only until lunchtime and would then be able to return home to eat lunch with Betty and then take an afternoon hike or walk together. Now that Mary-Anne works through the day and Betty is uninterested in walking alone, they rarely do any form of physical activity. Mary-Anne is concerned for both of their physical health due to the lack of daily exercise.

Solutions: The example of Mary-Anne and Betty highlights a number or issues.

- Even where there is strong desire to do something, circumstances may work against a person. As with several of our personas and scenarios, it is not just a matter of cognitive decline, but how that that plays out in everyday life

- Particularly, financial issues underlie a lot of the problems that people face when they get older and affordability needs to be considered in technology development

- Design may not just require us to visualize a single person. We may need to think about how the technology would be used by a couple, or multiple persons, together.

A potential solution is to provide low-income older adults with an exercise "prescription" as in one city in Finland. Older adults in a Finnish city get vouchers ($1,000+ per year) they can use to get free access to a gym with a trainer who specialises in fitness for older adults. This benefits them—they are healthier—and it is also seen as reducing demand on the care system. But it's not just about fitness—people feel better and more socially connected. The social/enjoyment dimension is crucial to the success of these schemes (Benke, 2020). Technology-based solutions include "exergames," which uses video gaming for training, education, sports, and health. Exergames are becoming more prominent in the field of health promotion and active aging, with increasing evidence of their long-term benefits, such as falls prevention (Ogonowski et al., 2016; Vaziri et al., 2016; Uzor and Baillie, 2019).

8.4   TECHNOLOGY FOR SUPPORTING HEALTHY LIFESTYLES

Exercise and activity are important components for maintaining everyday health and has positive benefits for maintaining cognitive health. However, for older adults and those with dementia, there can be barriers, perceived or real, to participating in regular physical activity.

This hitherto neglected area has increasingly become the focus for technology development in the AgeTech sector and some examples are outlined in the following sections. Investigators at Sheffield Hallam University, UK, are in the research and design phase of creating a social assistance robot that can be used in the maintaining safe exercise regimes for persons with dementia (Cooper, Procter, and Penders, 2016). Similar research examples can be found in other parts of Europe (Cooney et al., 2019), as well as the U.S. (Schrum, Park, and Howard, 2019), highlighting that finding methods to support and engage older adults with dementia in physical activity is a priority. Commonly among these research projects, the investigators recognize the importance of end-user collaboration for the visual acceptability, safety concerns, and overall functional features. More simplified technological innovations are being researched to aid in monitoring energy expenditure and activity to potentially lead to better person-centered physical health management. Using wearable fitness trackers, researchers in the UK have explored the relationship between energy intact and patterns of activity in persons with dementia in care homes (Murphy, Holmes, and Brooks, 2017).

Some older adults will experience physical injuries in their later life and will require therapy to help re-establish function and motor abilities. Investigators collaborating through Western University, in London, Canada, and the University of Waterloo, as well as Université du Québec, are developing technologies that can be used in various rehabilitation settings. They aim to develop technologies that can be used both at home and in a community care setting to provide individualized and tailored exercise programs and rehabilitation regimes. Ideally, these technologies could follow two separate approaches, one for less intense at-home use and another for more intense use under the guidance of a therapist (Jog et al., 2015–2020).

The role and availability of technologies to support the cognitive and mental health of individuals, both for caregivers and older adults, is growing. A collaborative research initiative out of Toronto, Canada, is exploring the use of information communication technologies to provide community dwelling for older adults with access to information to manage stressors of caregiving—such as providing apps/tools to manage symptoms of anxiety and depression. Furthermore, their app suite may also include mobile games intended to support and practice cognitive skills (Chignell and Liu, 2020).

As well as research projects, there are community initiatives and commercial products coming on to the market to support healthy lifestyles.

BOX 8.1 Dementia Friendly Sport and Physical Activity Guide

Developed by the Alzheimer's Society and Sport England, this guide is a downloadable digital resource that outlines guidelines and recommendations for safe sport and physical activity for persons with dementia (Bould, McFadyen, and Thomas, 2019). It a resource for individuals looking to support a loved one in physical activity, but is mainly targeted at leisure centers, gyms, and sports clubs to implement actionable changes to their programs and facilities.

Glowonconcept: Shutterstock

BOX 8.2 Rendever

Rendever is a virtual reality platform that can be used to support the cognitive and mental health of older adults (Rendever, 2019). Using a virtual reality system, Rendever can be customized to create a variety of experiences for older adults to participate in. Some of the key features Rendever offers are: (1) reminiscence therapy through the recreations of memories from the past; (2) opportunities to virtually leave their home setting to travel and cross items off their bucket list; and (3) create friendships through shared experiences, as many individuals can don their own headset and share the same virtual reality experience.

BOX 8.3 "Play Parks" and Recreational Space

A low-tech solution that is available to support the physical activity and health of many community-dwelling older adults is the implementation of free and accessible recreational spaces. Senior Play Parks (The Lindsay Advocate, 2019) are designed specifically for older adults as a facility to maintain mobility, flexibility, and agility, using low-impact, gentle exercise equipment. The intention is that anyone will be able to access these facilities regardless of income, ability, or other barriers.

8.5   KEY INITIATIVE—INDUCT PROJECT

Supporting healthy lifestyles through the use of technology is a broad area that has many potential applications and utilizations. Healthy lifestyles are not just limited to supporting physical activity, but also include important sectors such as nutrition and mental health. INDUCT (Interdisciplinary Network for Dementia Using Current Technology) is a research network in the European Union that is aimed at developing a multi-disciplinary collaborative framework in Europe to improve technology in order to improve the everyday lives of people living with dementia (INDUCT, n.d.). INDUCT has a number of research projects that fall under three main categories: (1) technology in everyday life; (2) technology to promote meaningful activities; and (3) health care technology. Within these categories are a number of specific research projects and initiatives that can be applied to promoting and supporting healthy lifestyles for older adults, who are experiencing cognitive declines. Some examples of projects include: evaluating the usefulness and effectiveness of "exergaming" technology available for older adults with dementia; evaluating how the use of digital arts and crafts can improve life satisfaction and quality of life; and exploring how Cognitive Stimulation Therapy can be converted and applied to tablet systems. These three example projects highlight the commitment of the INDUCT to having a holistic approach to supporting all aspects of personal healthy lifestyle factors for persons with dementia.

8.6   FIND OUT MORE

Global strategy and action plan on ageing and health—provides a political mandate for action required to ensure everyone's opportunities to experience a long and healthy life—A report from the WHO (2017): https://www.who.int/ageing/WHO-GSAP-2017.pdf?ua=1.

10 Priorities for a Decade of Healthy Ageing—Identifies actions for attaining objectives of global strategies and an action plan on aging and health developed by WHO International (2020): http://www.who.int/ageing/10-priorities/en/.

CHAPTER 9

Staying Connected

9.1 THE CHALLENGE

Social participation is about being able to remain actively engaged within society, the economy, and within our families and communities (Woolrych et al., 2019). Being involved in volunteering, work, leisure, religious and cultural activities, as well as civic engagement may improve the well-being of older adults and increase the social and human capacity of their communities. However, social participation often decreases with age due to illness and disability, loss of social contacts, or fear of rejection or discrimination. Social isolation and loneliness at all ages is a key issue in modern society. Older adults are particularly at risk due to living alone, health problems and disability, sensory impairment, and significant life events such as the death of a spouse (The Campaign to End Loneliness, n.d.). Reduced contact with family or friends may lead to adverse physical and emotional, health, and well-being outcomes (Smith and Victor, 2019). Initiatives to enable older adults to be well connected to their families, friends, neighbors, and communities will help older adults enjoy a healthier old age and remain actively participating in society. A number of technology-based solutions have been (and are currently being) developed to improve connectedness, including telepresence robots, systems to plan and support family caregiving, and social gaming. There is increasing evidence that technologies of various kinds can help to alleviate social isolation and loneliness.

Monkey Business Images: Shutterstock

However, the evidence base is still weak and further research is needed to demonstrate the benefits, usability, and acceptability of technology-based solutions (Khosravi, Rezvani, and Wiewiora, 2016). While social isolation is a problem for many sectors of society, the issue is likely to be especially acute for people with dementia and their caregivers. As we have seen earlier, cognitive health may be sustained through physical health, activity, and social participation and these may be compromised for those people who are socially isolated.

Crucial to supporting the continued social participation of older adults generally, and those experiencing cognitive decline or dementia, is the need to overcome the *digital divide* that exists within most societies, especially for many people who are marginalized. This is especially important for technologists and designers alike given that the design and operation of technologies are not typically friendly for older adults. It also requires us to promote equitable access to technologies (Vaportzis, Clausen, and Gow, 2017) through the principles of Universal Design. Access to broadband Internet and mobile networks are also important to allow older adults to remain socially connected and this requires attention to social, health, and financial equity.

9.2   WHAT'S IN THIS CHAPTER?

Understanding how we can develop technologies to best support older adults, particularly those living with dementia, is a complex undertaking. Making sure that such technologies have the intended positive impacts on health and well-being and avoiding any unintended negative impacts is important. To do this, technology development needs to progress with a good understanding of older adults' lives as well as those living with dementia. Below, the use of Albert's persona and scenario sets the scene regarding the complexity and shows how technological solutions can help. A range of useful technologies for connectedness are presented, including: the FamliNet platform (Box 9.1) and the ElliQ robot (Box 9.2) alongside gaming, digital storytelling, and other entertainment platforms. Finally, the Talking Mats framework is introduced as a resource for people with dementia to use in order to have their voices heard, make new connections, and actively participate in various conversation topics.

9.3   PERSONA AND SCENARIO—ALBERT

Persona: Albert lives in his own apartment in Sydney, Australia, built off the main family house of his only daughter, Liz. He was originally born and raised in England but after the death of his wife ten years ago he decided to move to be closer to his remaining family. Albert has made a few new friends in his time living in Australia, through his golfing league and a walking group in the neighborhood. Albert has been largely physically active for his entire life and has been an avid golfer since his retirement. He has always tried to maintain a healthy lifestyle as heart disease runs in his family. Up until four years ago, Albert never had any serious medical complaints, but he recently was finding that he had become more forgetful and sometimes struggled with following the pre-planned walking paths of his walking-group. Shortly after his initial complaints he was diagnosed with Alzheimer's disease that has been slowly progressing to more extreme and frequent symptoms. Albert used to enjoy his daily neighborhood walks, but has become frustrated that he cannot remember new routes on his own. When he first moved to Australia he would look after his two grandchildren when they came home from school, but the children are now older and do not need his supervision. Golfing has also become increasingly frustrating for Albert, as he has forgotten his tee-off times and his equipment on multiple occasions. The daily leisure activities and social connections he used to enjoy no longer hold joy for Albert.

Miljan Zivkovic: Shutterstock

Scenario: Although his dementia has progressed in recent years, Albert can still do many activities of daily living for himself, but often spends the entire day sitting watching TV, feeling unmotivated by his current life. Liz thinks this is bad for her father and that he will just deteriorate more if this lifestyle continues. A community health worker comes to visit Albert for an hour each morning during the week, but does not have much time beyond the basics of getting him out of bed, washed, and fed. Liz would like Albert to be more socially and physically active; she tries to visit with him as often as she can, but between work and supporting her two children with their activities, she finds this difficult.

Solutions: Recently, Liz discovered a pilot study for an interactive telepresence robot to aid older adults in stimulating communication and suggesting activities. After discussing the situation with her father, they both agreed that it could not hurt to take part in the study. Albert and Liz were given training by the study team on how to operate and utilize the features of Albert's new robot companion. Within the first week of set-up Albert has learned how to send and receive picture messages. He now enjoys exchanging pictures with his grandsons about their day. Albert often has chats with his robot, who he has affectionately named Jeeves. Liz is happy to find that Jeeves has been encouraging and reminding Albert to take a walk around the house, and plays some of Albert's favorite music while he does. Liz is hopeful that with the verbal prompts of Jeeves and the ease of communicating through voice and picture messaging, her father will feel less isolated when she cannot be there in person.

9.4   TECHNOLOGY FOR STAYING CONNECTED

The use of information and communication technologies (ICTs), mobile technologies, wireless networks, and the ubiquitous Internet has huge potential for helping older adults stay connected. Areas of research are exploring the use of technology for supporting cognitive health through social means. By facilitating the social needs of individuals at any stage of cognitive health, individuals may be better supported in their pursuit of vocational and leisure activities, as well as overall well-being. The development of new technologies and better understanding of how existing technology can be used for social connectedness while ensuring that virtual social connectedness does not replace person-to-person contact is a growing area of research.

Technologies to facilitate connection between people with dementia and their families, as well as the wider world, is now emerging as an important line of research. For example, Ticket to Talk is a digital intervention that facilitates the exchange of personal memories and inter-generational conversation (Welsh et al., 2018). Virtual reality can also deliver experiences that are no longer attainable by providing a creative medium for comfortable and enriching experiences that can promote and improve the quality of life for PwD (Hodge et al., 2019). Foley and colleagues (2019), used tangible objects as moments of (mutual) recognition, and meaning to provide opportunities for collaboration amongst residents in care homes. Similarly, the use of exergames in a care home were used to foster collaboration and social experiences within a group setting (Unbehaun et al., 2018). Participation and being involved in a conversation can create a sense of belonging and self-worth for PwD.

A joint research project through the University of Toronto and Sunnybrook Hospital, Toronto, Canada aims to investigate the use of web-based connections for minimizing social isolation. Their goal is to design a web-based platform to assist older adults to interact, in a user-friendly format, with family, friends, caregivers, and organizations in order to improve social support and connectedness. As approximately 25% of older adults living at home in their wider community in Canada express some feeling of social isolation (The National Seniors Council, 2014), designing tools to help alleviate these concerns, and the associated negative mental health consequences, is critical.

Gaming and other entertainment platforms have a role in promoting physical and cognitive activity and leisure, but they can also have an important social dimension. A research group at

Simon Fraser University (SFU), Canada, is exploring how digital games for learning and entertainment could be used to increase social connectedness. Specifically, they aim to design, create, and test digital game platforms with a series of games designed for older adults. Their research has explored the unique use of digital gaming and how it can impact socioemotional connections (Zhang and Kaufman, 2015), and be used to create intergenerational play (Zhang et al., 2017). Their results indicated that the use of digital gaming can be used as a platform to engage younger and older adults in meaningful interactions and bridge intergenerational gaps by fostering social gameplay.

The use of digital storytelling is an expansion of the age-old practice of telling stories by using technology—in various media forms such as text, photos, sounds, and video—to reimage a story, pass down personal history, and share experiences (Rule, 2010). A growing research body is exploring the benefits of older adults using digital storytelling for social connectedness and general well-being. Although research about older adults using digital storytelling is fairly novel, there are many expected benefits for using digital storytelling. Digital storytelling can create opportunities for older adults to reflect on their life histories and share their narrative, building new connections and relationships (Waycott et al., 2013). Additionally, digital storytelling helps to interconnect generations, a personal digital story could be passed down between generations and retold many times (Hausknecht, Vanchu-Orosco, and Kaufman, 2019).

AgeTech for connecting people, enhancing social interaction, and helping avoid social isolation is also a focus for community initiatives and commercial products and services.

BOX 9.1 FamliNet

FamliNet (https://www.famlinet.com/) has developed a connections-communication platform that helps prevent loneliness and isolation by keeping older adults in contact with family and friends (AGE-WELL News, 2020c). Unlike other traditional communications platforms, FamiNet has been designed specifically to account for challenges older adults may face such as cognitive difficulties and language barriers, by including features like speech to text, audio and video recording, and automatic translation. It is an easy-to-learn platform that features pictures of the user's contacts and very simple icons to access the different types of messaging: text, pictures, and video. As a custom Web app, it allows an individual to stay connected with family and friends using the Internet on a secure data server.

Figure 9.1: Stephanie Gagne and her father Richard Ratcliffe are regular users of the FamliNet communications platform. Says Gagne: "It opened up horizons for my dad. He's now challenged and stimulated in a way that he hadn't been for a long time." Photo by John Hryniuk, courtesy: AGE-WELL NCE.

BOX 9.2 ElliQ

ElliQ is a fixed position robot, a telepresence, which utilizes artificial intelligence to gauge and learn what tasks and interactions the user may like and need. ElliQ was designed specifically with older adults in mind, boasting a number of features, such as voice responsive prompts and cues, personalized activity suggestions, messaging and video calls, photo-sharing, and event reminders. ElliQ uses both a visual interface, through a tablet screen, and body language. ElliQ will move and respond to the user's voice, gaze, and touch to communicate at a deeper level than the average voice assistant tool. As ElliQ collects data suggestions will grow and become better tailored to the user's preferences and wants. ElliQ helps create social connections between loved ones by using a private app that shows when users are online in order to utilize the messaging functions.

9.5   KEY INITIATIVE—THE TALKING MATS

The deterioration of communication abilities is a major concern and often a distressing aspect of dementia. With the decline of an individual's communication abilities it becomes increasingly more difficult to ensure that their opinions, views, wants, and desires are understood and heard. The Talking Mats framework has been a critical initiative in enabling people with dementia to have their voices heard. The Talking Mats are a communication framework that uses picture symbols in order to create a practical communication system that assists in an individual expressing both positive and negative views. It is available in both a low-tech version and a digital tablet format. The availability of both options allows for both accessibility in terms of cost, but also considers the potential of the digital divide where some individuals may not be receptive to a high-tech system. Studies evaluating the use of The Talking Mats have found that, for people with dementia, The Talking Mats are effective for promoting engagement in conversations, increasing the reliability of information provided, and increasing overall conversation time (Murphy et al., 2010). The Talking Mats framework has been successfully used over the past decade and continues to be a useful resource in supporting the communication needs of people with dementia.

9.6   FIND OUT MORE

Baecker, R. and Black, S.—Promoting social connectedness through new and innovative communication platforms-connect-tech: https://agewell-nce.ca/research/research-themes-and-projects/wp4-new.

Effects of technological innovations on social interactions, an article by Kim Boucher Morin (2018)—examines social isolation and technology. It discusses how technology can be used to reduce social isolation among older adults: https://www.socialconnectedness.org/effects-of-technological-innovations-on-social-interactions/.

Kaufman, D., Sauvé, L., and Ireland, A. (2020). Playful aging: digital games for older adults.—Is a white paper by the AGE-WELL 4.2 project February, 2020: https://agewell-nce.ca/wp-content/uploads/2020/02/AGE-WELL_WP4.2_White-paper_GAMES.pdf.

World development report, 2016: Digital dividends. World Bank Group—This report, edited by Bruce Ross-Larson, focuses on how to make digital technologies beneficial for everyone by closing the digital divide, with a focus on Internet access: http://www.worldbank.org/en/publication/wdr2016.

CHAPTER 10

Financial Wellness and Employment

10.1 THE CHALLENGE

AgeTech is very much about supporting the well-being and health of people as they grow older. Obviously, we hear a lot about the physical and mental health of people, but a person's financial wellness is about the overall financial health of a person, sometimes referred to as financial security. This is an important component of healthy aging (Sixsmith et al., 2014) and despite the image of wealthy baby boomers, many people face financial hardship as they grow older. Low income has implications for health and social engagement, and it increases the risk for major problems such as homelessness. At the societal level, population aging has financial and social implications for the sustainability of public welfare systems throughout the world. How can governments make best use of available financial resources to ensure that older adults continue to have a decent income and quality of life?

This is still a relatively new topic within the AgeTech sector, and the key priorities for research and development are still emerging, and this is especially the case in relation to people with dementia and in relation to cognitive health more generally. In this chapter, we try to capture some of the issues, but this remains very much a work in progress.

One of the major issues facing people living with dementia in terms of financial wellness concerns the heavy economic outlays for care provision, with some research indicating that families pay more for the care of older adults than their children (Kim and Antonopoulos, 2011). Given this, there are substantial economic gains to be made if cognitive health can be protected via the reduction of acute care costs and the maintenance of occupational productivity from both informal caregivers and the individual experiencing cognitive health decline.

Another area that has emerged as important is how to support people to remain in the workforce as they experience age-related cognitive decline and mild cognitive impairment (MCI), as this will contribute to their financial well-being. Age-friendly workplaces as well as training and education

Wavebreakmedia: Shutterstock

initiatives will help people to remain in the workforce, and support companies and organizations in recruitment and retention of workers in key areas such as health care and other labor-intensive industries (Nagarajan et al., 2019). However, many individuals who experience cognitive decline are unable to participate as actively in the economic market as they did previously, whether this is through occupational work, volunteering, travel and leisure, or other spending habits. Technology, both developed and developing, in this challenge area includes workplace technologies to support aging workers, mobile financial apps, cyber-security programs, age friendly banking, and technology-based training and education.

10.2   WHAT'S IN THIS CHAPTER?

Financial wellness, as a component of healthy aging, affects many areas of a person's life, as shown in Edith's persona and scenario. A range of solutions can be put in place for people like Edith, with technological solutions to provide more supportive financial and workforce supports. Examples of research underway in this area are outlined in a section on technology for financial wellness and employment. Box 9.1 gives an example of a free targeted online course on mobile banking. Box 9.2 covers technology that helps older adults avoid/deal with online scams. The chapter ends with a brief description of promising initiatives from Canada designed to promote labor force participation of older adults (Government of Canada, 2018).

10.3   PERSONA AND SCENARIO—EDITH

Persona: Edith is in her early 60s, lives in Toronto, Canada, and is starting to think about her eventual retirement from her career. She has worked in the financial services sector for most of her career and she has seen a lot of changes, many of which she doesn't think are for the better. She doesn't like the workplace culture anymore and her job is very different from what it was years ago. She has a good retirement plan and plenty of savings and could take early retirement, but is still hesitant about leaving the working sector. Edith is in general good health without any major concerns. Her only complaint is that she finds some aspects of the job take her a lot longer than they used to and this makes her feel anxious. Her husband Stu is already retired and can often be found babysitting their grandchildren. He has been talking more frequently about traveling for enjoyment. However, Edith is not so sure. She has always been a self-declared workaholic and she is worried about what she will do with so much time on her hands. She also thinks she still has a lot to offer in the labor market.

Paulo Vilela: Shutterstock

Scenario: Edith has experienced a lot of changes in the workplace. Technology has changed the job beyond recognition from when she first started and recently she has found it difficult to keep up with all the changes. The company accounting system and intranet was recently upgraded and it took her a long time to get used to things—a lot longer than her younger colleagues who seemed to have no problems. She was shown what to do several times by a colleague, but still forgot things. She feels embarrassed asking people for help again and tries to avoid using the system as much as possible. Because she has found it hard to adapt to the technology change, she is worried that her new (much younger) boss will think she is ineffective, and she increasingly feels she is being left behind. She thinks she perhaps should get some training to upgrade her information technology (IT) skills but lacks the confidence and motivation to "change again at this time in life."

Solutions: The example of Edith highlights the situation facing many more people as the world's population ages. Perhaps both society and older individuals need to evolve new ways of living in later years. For example, in Japan, a country with one of the oldest populations in the world, there is an initiative supported by the government called the "Second lives scheme." It combines employment opportunities with supportive community design. When we think about "supportive communities," we should think about older adults being part of, and contributing to, their communities. They are not just passive recipients of "support." Supportive technologies needs to be built into these solutions in ways that will enable continued social and economic participation and not marginalize people within the workforce.

10.4   TECHNOLOGY FOR FINANCIAL WELLNESS AND EMPLOYMENT

The practical use of technology to support older adults in their financial wellness and employment is now emerging as a major focus of research and development in AgeTech. For example, part of AGE-WELL NCE's research program is directly focused on this challenge area (AGE-WELL NCE, 2020). Examples of these research projects include: (1) an employee-employer database system that can minimize employment barriers by matching employees to employers; (2) a collaborative stakeholder project to investigate how workspaces can become more accessible for persons experiencing mild cognitive decline or early stage dementia; and (3) a project that models sustainable social enterprises that will offer employment to people with cognitive disabilities to reduce their digital exclusion.

Workplace trends across many sectors are seeing two major shifts: (1) a larger number of older adults transitioning out of the workforce; and (2) technology use within jobs is becoming more prominent through increased automation and digitalization. Providing training support for new technology use is important for older adults to remain or re-enter the workforce (Federal/Provincial/Territorial Ministers Responsible for Seniors Forum, 2012). Evidence shows that using an e-learning training program, developed for older job-seeking adults lacking the requisite technology skills to find employment, is effective and feasible for building technology skills for older adults (Taha, Czaja, and Shari, 2015). Researchers in the U.S. and Portugal are investigating the usability of various types of wearable trackers to manage and tackle the challenges of the aging workforce such as training of new workers and monitoring vital signs during physical labor. Through tracking specific parameters, the researchers state that there are potential implications for which the data can be used for health and safety and training. For example, they discuss the use of augmented reality to record job tasks from the first-person perspective to enhance training (Lavalliere et al., 2016). Some recent examples of community initiatives and commercial products include supports for mobile banking and avoiding online scams.

BOX 10.1 "Ready, Set, Bank"

Sponsored by Capital One Financial Corporation, "Ready, Set, Bank" is a free online course specifically designed for older adults to learn about the basic in-and-outs of mobile banking and to gain the confidence to use online banking tools. The online course features lessons on the benefits of online banking, safety and security, and how to get started. The course also includes many demonstrations on how to use specific features like paying bills, depositing checks, and transferring funds.

BOX 10.2 CPR Telephone Blocker

Unfortunately, many older adults can fall prey to malicious phone scammers causing both financial, as well as personal and privacy losses. The CPR telephone blocker is designed to remove the ability of robo-calls and telemarketers of contacting the landline it is attached to. It has the ability to simply manage the numbers that are on the block and allow list. Depending on the level of security desired, the CPR blocker can range in function from allowing numbers to be blocked in real time by the user, or be set so that only specific numbers will be let through and all others blocked. The device provides a large display to indicate if the number has been blocked and how many times it has tried to call since being blocked. Additionally, a mobile version of the CPR blocker also exists for people who do not have a landline or have a mobile phone in addition to a landline.

10.5   KEY INITIATIVES

There are a number of promising initiatives that may be useful in promoting the labor force participation, including the use of technology. However, there is still limited evidence within Canada as to what would be considered *good practice* for using technology to support older adults' workforce and investigation into the transfer of these initiatives into an AgeTech context is an area that needs more work. Some of the promising initiatives identified are described below.

- **Initiatives that lead to more flexible work for older adults**—Flexible working conditions and hours is an area of occupational health that has been shown to be related to the well-being and longevity of the working lives of older employees (Göbel and Zwick, 2010). This practice has existed in the UK since 2014 and has seen positive results since its legislation. Flexible work arrangements serve as a means to better balance work–life commitments.

- **Initiatives that lead to modifying work environment and tasks**—Modifying work environments is an additional initiative that can assist older workers to remain in their working capacity longer (Göbel and Zwick, 2010). As older adults vary in the need for work supports, this initiative is inherently more complex. Modifications in working conditions can vary from adjusting lighting, providing ergonomic accommodations, reducing environmental noise or adaptable work equipment to name a few examples. The cost and complexity of what kind of technology could be utilized to modify specific working conditions is an area of research that should continue to be explored.

- **Collaborative education providers for older adult learners**—Older adults looking to remain in the workforce or to re-enter the workforce may often need training and education to better understand the changing demands and nature of the work. By strengthening the relationship between employers and learning providers the design and accessibility of training programs for older adults could be better delivered.

Understanding the barriers of educational training for older adults is the first step in ensuring that new skills training is accessible to older adults.

- **Job-matching programs that specifically target older adults**—Job-matching programs are one of the most common initiatives implemented internationally, many of these initiatives help to balance the need for skilled workers while taking into the consideration of the special needs of the older adult job seeker. Currently the Canadian website "Job Bank" allows the functionality for employers to indicate their willingness to hire older workers to fill their vacancy, through the use of a filtering option these jobs can be explicitly found.

10.6   FIND OUT MORE

World Economic Forum (October 5, 2015)—A white paper that discusses how 21st century longevity can create markets and drive economic growth. Cologny/Geneva: World Economic Forum: http://www3.weforum.org/docs/WEF_GAC_Ageing_White_Paper.pdf.

Taylor, M. T. and Bailey Bisson, J. (2019)—Changes in cognitive function: practical and theoretical considerations for training the aging workforce. *Human Resource Management Review*, 30(2): https://www.sciencedirect.com/science/article/pii/ S1053482218302535.

Longevity and pension plan sustainability: the rising proportion of the population aged 65 and above leads to a longer payment period for pension—http://www.acpm.com/ACPM/ The-Observer/Fall-2016/Industry-Insider/ISSUES-ANSWERS/Longevity-and-pension-plansustainabilityaspx.

CHAPTER 11

Co-Creating Technologies with People Experiencing Cognitive Decline

11.1 INTRODUCTION

Increasing emphasis is being placed on research that involves co-creation (also co-production, participatory action research (PAR), etc.), approaches, and methods. Co-creation (Battersby et al., 2017) is about involving customers, end-users, and other stakeholders in stages in the development of a new product or service. These stakeholders are not merely providers of information—they are also active contributors and collaborators within a research team. Most importantly, they are the experts on their own experiences and situations. In this chapter, some of the general principles and methods of co-creation are outlined, followed by specific approaches for working with people with dementia, as this group presents a range of additional challenges.

Monkey Business Images: Shutterstock

Some groups, such as people experiencing cognitive decline and/or dementia, are often excluded from the design and development process, despite the increasing emphasis on co-design approaches in both research and industry (Brooks, Savitch, and Gridley, 2017). This has meant that the very people for whom the technology is being developed for are not part of the process of creating ideas, solutions, and interventions that will help them to age well at home and in their communities (Brooks et al., 2017). There is also a growing recognition of the value of grounding research in lived experiences to better ensure useful outcomes and relevant outputs, and in shaping the design and development elements of research (McKeown et al., 2010). It is important that people experiencing cognitive decline be involved as active, contributing members of the team and not just passive research subjects (Tanner, 2012). However, their role in current design practice is often viewed as "end-users" and research subjects, resulting in their superficial involvement on projects (e.g., as part of a nominal advisory group alongside using their caregivers as "proxy" informants). Yet, it is important to be aware that there are methodological and practical challenges and legitimate ethical reasons why researchers might be unable to co-produce with people with severe dementia.

Not surprisingly, the question of "how can we better *work with*" people experiencing cognitive decline has yet to be answered. This chapter provides some ideas on how improvement can be made in the ways researchers and developers work *with* people experiencing cognitive decline to co-create solutions. Key points of focus for this chapter will include: an overview of co-creation ideas and good practices for working with people experiencing cognitive decline; a step-by-step breakdown of ways to co-create; and this is followed by a discussion of some of the challenges. It is not possible to look at co-creation methods in too much depth in this chapter, but this chapter has tried to provide some general principles, with particular reference to people with dementia. Readers are encouraged to follow-up in more detail with the learning resources listed at the end of the chapter.

11.2   KEY PRINCIPLES IN CO-CREATION

It is important to outline some basic principles of what a research partnership could look like. Community-based participatory research approach (CBPR) is a valuable approach that has inspired researchers, designers, people who live and work in the community, and industry professionals, alike. This approach can help individuals to develop more equal partnerships and co-create solutions to improve peoples' health (Jagosh et al., 2015) and has become increasingly used across academic disciplines, government and non-government sectors, and in other humanitarian work areas (Jagosh et al., 2015). Key ideas and ways of working using CBPR are outlined in Table 11.1.

Table 11.1: Key ideas and ways of working using CBPR

Idea	Purpose	Action
Equity	To ensure a balance of power among different groups of stakeholders.	Create a group with a range of people who have a vested interest or who are beneficiaries of the end result at the start of the project to help create the solution or intervention and be involved in the decision-making.
Inclusivity	To make the most of opportunities for all people involved to participate in the research, planning, and development process.	Develop opportunities for meaningful exchange of knowledge using less conventional ways of working (e.g., mapping workshops, informal dialogue sessions, knowledge cafés) to enable participation of people with varying degrees of knowledge, expertise, and skills in research.
Empowerment	To provide people who are most affected with the opportunities and resources that will help them to be more confident to take action in the research and design process and in the decision making.	Implement ways to more easily join meetings, collect, analyze, and share ideas, data, and solutions. Communicate often with people who are less vocal and share new information using easy to digest mediums, e.g., telephone and Facebook (has become a popular social media platform for people (55+) and for those living with mental health conditions).
Partnership	To work with a range of stakeholders (i.e., different people who have a vested interest in the idea or intervention such as persons experiencing cognitive decline or informal caregivers) as partners and work toward a creating a shared goal, ideas, ways of working, and solutions.	Create a list of stakeholders at the start of the project and look to see if anyone is missing from the table. Meet with them to agree on a way of working, role of partners, ways to communicate. Include partners in all stages of the research and design process and ensure that people who are less vocal have their voices heard either during discussions or through one-to-one follow-up meetings.

| Co-creation | To develop new knowledge and solutions to complex problems together with diverse stakeholders. | Use less conventional ways of working (e.g., creative methods such as photography and storytelling) to generate new knowledge with people who may not have research skills. Make the best use of the skills that they have. Prioritize using their everyday experiences as "evidence" to inform decision-making in the research and/or design project. |

Source: Adapted from Jagosh et al. (2015).

11.3   CO-CREATION: A STEP-WISE PROCESS

As already mentioned, co-creation can be defined as a collaborative process, whereby the people who are going to use a new device, product, or service are actively involved in "creating" the knowledge or new technology. This kind of approach can be challenging and requires significant commitment and effort for everyone involved. However, there are potential benefits that make this effort worthwhile, as new solutions are grounded in the realities and experiences of those who are going to use them. This means the solutions are likely to be more meaningful, acceptable, and user-friendly, while design weaknesses can be identified early to avoid additional costs and time. This section outlines key steps to guide the co-creation process for co-research and design alongside case examples.

11.3.1   STEP 1: IDENTIFY STAKEHOLDERS

The key question here is: *Who needs to be at the table*? Stakeholders are those who also have vested interest, in the area being addressed, e.g., people experiencing cognitive decline, informal caregivers, service providers, and government. Different stakeholders need to be involved, as they provide their particular perspectives on a project. One systematic way of identifying stakeholders and their roles is to create a map or *Actor Model* (Lunn et al., 2003). This is usually a table that contains a complete list of stakeholder types, the specific stakeholder, their role and their particular goals/priorities. The model should be something that is jointly created and refined by the stakeholders involved so that there is buy-in from everyone concerned. It is likely to be incomplete but will serve as a basis for bringing together an initial stakeholder group. Discussing the model with stakeholders will help expand the constituency of the group and should be revisited and refined as the research progresses.

11.3.2   STEP 2: RECRUITMENT AND AGETECH APPROPRIATION: ENGAGING PEOPLE EXPERIENCING COGNITIVE DECLINE AS CO-RESEARCHERS AND CO-CREATORS

This is not just about implementing a recruitment strategy. Hellström et al. (2007) describe having several *layers* of gatekeepers who must be consulted and negotiated with in order to gain access. These can include: health service providers; family members and carers; management (of institutions); and ethics board/committee members. Gatekeepers often have the decision-making power over the people they are *gatekeeping* and thus can decide *whether* they participate and/or *how* they engage in the research. Gatekeepers should be thought of as a stakeholder group, and it is important to include gatekeepers so that they are informed allies, rather than ill-informed adversaries (also see Chapter 12). Possible avenues to access both gatekeepers and persons with dementia are charitable organizations, dementia support groups, older adult day centers, community centers, health service boards, working groups and committees, churches, and research groups within academic institutions.

11.3.3   STEP 3: MAINTAINING ENGAGEMENT AND INVOLVEMENT

As stated by McKeown et al. (2010), "the notion of active involvement in the co-creation process rests on values of transparency, honesty and openness, perhaps then it is time to extend this to honesty about the work." Considerable effort is required by the project leads and other research and design team members to nurture and sustain good working relationships through frequent and open communication (Tanner, 2012). This can be carried out by developing a communication protocol with co-researchers (and other project stakeholders alike) at the start of the project (see Suter et al., 2009):

- establish a common language (e.g., protocol for querying and explaining dementia-related clinical terms during discussions and meetings);

- frequency of communication (e.g., weekly, monthly, quarterly);

- platform for remote, digital engagement (e.g., Skype, Zoom, WhatsApp);

- communication etiquette (e.g., only during regular business hours unless urgent);

- processes for conflict resolution (e.g., face-to-face discussion between persons in conflict, followed by escalation to higher-ups if necessary);

- channels to bridge connectivity (e.g., through administrative gatekeepers or directly to the individual);

- modes of communication (e.g., in-person, group meetings, one-on-one meetings, by phone, via remote platforms);

- use the most effective mechanisms for communicating with different stakeholders (acknowledging in different life circumstances); and

- develop personas and scenarios to facilitate empathy within the design team.

11.3.4 STEP 4: CAPACITY BUILDING AND TRAINING

Co-creators need to feel confident in their role and ability to contribute in a meaningful way to the research and design process and maximize the potential impact of project outcomes and outputs. Project leads need to organize and conduct (as soon as possible) informative and accessible training sessions (i.e., focused on participant recruitment, gaining informed consent, data collection and analysis, reporting, and dissemination) with co-researchers. This is the key for optimizing engagement and productivity, ensuring the best use of co-researcher skills and experiential knowledge. In addition, some mechanisms for upskilling and training can include: shadowing and participant observation; organizing topic-specific research and design workshops (e.g., building trust and rapport, recruitment, conducting interviews, ethical research practice, co-analysis with participants, reporting, and presenting at academic conferences); mock exercises; and keeping a reflexive journal to identify key challenges for discussion.

11.3.5 STEP 5: CO-DESIGNING AND DEVELOPMENT

Effective co-design requires the project team to work iteratively, creatively, and flexibly. Co-creation stages for consideration whereby people with dementia are closely involved in all phases.

- Identifying initial ideas and priorities using methods such as ideas/concept mapping, and informal, unstructured group dialogue (e.g., deliberative dialogue, world cafés). Deliberative dialogue refers to group discussion method aimed at generating thoughtful conversations, unique from other forms of public discourse techniques such as debating, negotiating, ideas mapping, and generating consensus. Key characteristics: multiple stakeholder participants; shared platform; informal; encourages ideas exchange and requires the generation of actionable tasks at the end of the dialogue session (Plamondon, Bottorff, and Cole, 2015). The World Café methodology generates in-depth discussions, by engaging a large group of cross-sectoral stakeholders in concurrent smaller group dialogues, in order to explore a single question or use a progressively deeper line of inquiry through several conversational rounds (Brown, Isaacs, and World Café Company, 2002):

- iterative development through a user-driven/person-centered approach of: concept development; story-boarding concepts; prototyping of products (by both persons with dementia and co-researchers); and

- piloting and evaluation at home and in the community.

The various phases of the iterative development process can be re-occurring should difficulties occur i.e., when issues are identified, the research and design team (inclusive of co-researchers and person with dementia) can go *back to the drawing board* carry out another iteration of: (1) meet and discuss; (2) co-analyze and report; and, (3) redevelop and implement. Figure 11.1 demonstrates the results of a workshop for the Age-Friendly Living Ecosystem project.

Figure 11.1: The graphic was drawn in real time while a full group discussion progressed around key themes and ideas at the first co-creation camp where we considered: What is an intergenerational age-friendly living ecosystem? And what does this look like? Courtesy: Clare Mills of Listen Think Draw (www.listenthinkdraw.co.uk)

Case example Box 11.1: Working with Older Adults: The Case of Affordable Housing

Researchers from SFU worked with an older adults' housing society and the municipal government to oversee the transition of low-income older adult tenants from old, dilapidated accommodations into an affordable housing redevelopment project. A CBPR approach was applied using a variety of methods (i.e., in-depth interviews, photo-voice, community mapping, deliberative dialogue, feedback forums, and knowledge cafés) to help not only identify key stakeholders but also co-create solutions to enable the successful transition of older adults into affordable housing. This research translated the lived experiences of low-income older adults and community stakeholders into formal and informal supports that fostered meaningful home environments for the older adult tenants, created a role for older adults as active *place–makers* in community planning and development. It has attracted media attention concerning the need for more affordable housing for independent living older adults (Fang et al., 2017).

11.4   APPROACHES AND METHODS FOR WORKING WITH PEOPLE WITH DEMENTIA

Involving people experiencing cognitive decline in the research and design process can be challenging (Hendriks et al., 2014) and, because of this, the research planning process tends to be *exclusive rather than inclusive* (Brooks et al., 2017). A review by Span et al. (2013) found that out of 26 projects that investigated the involvement of people experiencing cognitive decline in the research process when developing IT applications, only 2 had included them as a decision-making partner. Even where researchers and designers include people experiencing cognitive decline as informants or advisors, they are seldom part of the decision-making team. A more active involvement in all aspects of the research will ensure that the research and development is truly grounded in lived experience. For example, research has found that the focus of their involvement still tends to be directed at the difficulties relating to the symptoms of living with cognitive decline rather than the day-to-day experiences (Lindsay et al., 2012).

While it is still an emerging area, there has been growing interest in developing approaches, methods, and tools to facilitate co-design and participatory approaches that support and encourage the involvement of people with dementia in research; however, research on co-design and participatory approaches for co-developing AgeTech remains limited and often utilizes more general participatory ways of working with older adults. Technology co-design methods have largely stemmed from qualitative research and the majority have focused on working with individuals who have mild to moderate cognitive impairment and/or dementia. Working with people with advanced dementia is often avoided due to several ethical concerns and risks (see Chapter 12 on ethics). Therefore, the methods presented in this section are specifically focused on working with people living with mild to moderate dementia.

11.4.1 METHODS FOR CO-DESIGNING WITH PEOPLE EXPERIENCING MILD TO MODERATE DEMENTIA

There are various innovative ways to work with those experiencing mild to moderate dementia in qualitative research (Phillipson and Hammond, 2018) including: co-researching, visual methods, participatory methods, ethnography, and use of Talking Mats.

1. Co-researching: This way of working integrates people with dementia as part of the research team as trained researchers who, for example, will conduct interviews with participants also diagnosed with dementia or be involved in the analytical process. Co-researching is grounded on a participatory and emancipatory philosophy as part of action research, but has reportedly demonstrated more personal benefits for participants with dementia (Tanner, 2012).

2. Participatory and Visual Methods: A key part of co-design, these methods have been used to involve people with dementia in a more hands-on way that can include the use of visual methods such as photography or filmmaking (Capstick and Ludwin, 2015) in addition to writing and acting (Jenkins, Keyes, and Strange, 2016), diary keeping (Bartlett, 2012), design (Boman, Nygård, and Rosenberg, 2014), and autobiographical narrative production (Benbow and Kingston, 2014). Being more hands on in the research process was reported to have helped participants with dementia feel more involved and with the benefit of having co-produced something that is tangible in the real world.

3. Ethnography: Research has shown that the involvement of people with dementia can be enhanced by using techniques that support and extend memory and communication and with less reliance on memory recall. Ethnography can enable "in situ conversations" to take place to help participants with dementia experience research events in real time. In one research project, these events were also filmed to capture practices and interactions between people with dementia, people with dementia and the physical environment, and people without dementia who participated in the study (Ward and Campbell, 2013).

4. The Talking Mats: This method uses a semistructured framework that applies a visual scale to acquire views on a range of research topics and have been shown to enhance effective communication, especially for those in the later stages of dementia. It was developed as a visual tool as a response to the difficulties linked to deteriorating verbal communication and comprehension for people with dementia to help promote self-expression when verbal abilities are compromised (Murphy et al., 2010).

11.4.2 PRACTICAL INITIATIVES THAT INVOLVE PEOPLE WITH DEMENTIA IN RESEARCH AND DESIGN

A systematic review by Suijkerbuijk et al. (2019) explored human-centered design projects that focused on people with dementia. The review highlighted the value of co-design approaches, reporting that people with dementia can influence technology development particularly in regards to content, design, and even the initial idea. However, there is still a lack of appropriate methods and materials for active involvement of people with dementia. The boxes below highlight three initiatives that provide useful guides to how to practically involve people with dementia in research and design.

Box 11.2 KITE (Keeping in Touch Everyday)

Developed by Robinson et al. (2009, p. 522), KITE aims "to foster an empathic relationship between designers and people with dementia by demanding close, respectful contact," while focusing "on trying to develop a holistic understanding of people with dementia's day-to-day lives." The method comprised a three-stage participatory design process: scoping stage involving focus groups; design stage involving co-design workshops; prototype stage involving try-out meetings. This approach afforded insights into their everyday realities, empowering people with dementia through meaningful engagement in the research and design process leading to design outputs, including in-car reminder technology and a wearable device to help joggers and runners. The study showed that involving older adults in the process of participatory design is feasible and could lead to devices which are more acceptable and relevant to their needs.

Box 11.3 OA-INVOLVE

OA-INVOLVE is a core AGE-WELL NCE project focused on researching practices and experiences of older adults' engagement in technology projects. The goal of the project is to develop ways to promote their active and meaningful engagement in the entirety of the research process, and facilitate their participation. One of the guides developed by the OA-INVOLVE team provides an effective research strategy and practical tips for engaging older adults living with dementia in technology development.

➢ Support the autonomy of people living with dementia in the research process but when necessary invite a family member (caregiver) to the meetings if requested by the older person.

➢ Connect with gatekeepers from community-based organizations (e.g., dementia care services), faith groups, memory clinics, and caregiver support groups to assist with recruiting people living with dementia.

➢ Create place where people feel physically, mentally, and emotionally safe. Establish what this space means for people living with dementia and their caregivers during initial project planning activities, and then verify it during subsequent meetings.

➢ When introducing a project to people living with dementia, do so in a place of their choosing and their familiar surroundings.

➢ Support the well-being of people living with dementia by asking them (or caregivers) how to best ensure their comfort; ensuring they arrive home safe after meetings and having their emergency contact information at hand.

➢ Obtain ethical clearance early on and invite dementia experts (e.g., service providers, frontline workers, and persons with dementia) to review the ethical protocol.

Box 11.4 Tungsten Tools

Tungsten (Tools for user needs gathering to support technology engagement) developed by Astell and Fels (2015–2020), is an AGE-WELL NCE project that has developed a set of tools to involve older adults in the early stages of design and testing of products and to provide ideas for marketing and advertising. The tools include:

- Technology Interaction—Learn how people adopt new technologies and discover potential barriers

- Show and Tell—What makes people choose or abandon technology products?

- Scavenger Hunt—Showcase your new products and improve it with real people.

A hands-on guide is available to help users create better technology for older adults. Tungsten tools can be found at: http://tungsten-training.com/tungsten-workshops

Based on a review of guidelines for designing state of the art technologies for people with cognitive impairment (Astell, Czarnuch, and Dove, 2018), there are three key areas for consideration: (1) cognitive; (2) physical; and (3) social.

1. First, the design plan of the product must consider how it can improve *cognition*. To do this, the goal of the product should be very clear and should aim to help the user maximize the ability to retain learned skills. The product should also be easy to use, be understandable, and consist of a clear set of how to use instructions with the effective use of user prompts and features that help to "gain" and "sustain" the individual's attention. Ideally, it is designed in such a way that avoids timed responses from the user and is fundamentally "failure free."

2. Second, it is important to think about the *physical challenges* that can occur when living with a cognitive impairment. Questions that a designer should ask are: Can this product help with mobility? Does it account for inexact motor control? Is it age appropriate and is it intuitive to use? Is it appropriate for people with various physical challenges? Can it be used by people with audio-visual impairment?

3. Third, when thinking about user adoption and uptake of the product, designers need to consider what would make their design *socially appropriate* for different groups. What sorts of features would help to spark interest for different people with cognitive impairment? To answer this question, it is important to think about iterative designing—meaning having timely user input throughout the design process. Equally, people like having positive timely feedback so positive reinforcement should be a key design feature for consideration. Also important for consideration are issues concerning privacy and ethical issues associated with co-designing and this is further discussed in Chapter 12.

When preparing for your next design, really think about these three areas. Try applying them to the persona and scenario below.

11.5   PERSONA AND SCENARIO—DOROTHY

Julie Campbell: Shutterstock

Persona: Dorothy is an 83-year-old widow who lives in a big two-level house in Vancouver, Canada. She lives with her 46-year-old daughter who has bipolar disorder and requires some care. Her husband recently passed away at age 92 and he had both Parkinson's and dementia. Although very active in her old age, in that she still walks to the grocery store every day and takes public transportation to her acupuncture appointments. Even though she has bad sciatica in her lower back, she still carries on with her daily routine and is coping with proper treatment physical and medicinal treatment. Some days are worse than others. To make matters worse, more recently, she noticed that her memory is getting worse and worries that she might not be able to continue with her routine of going to the grocery store every day.

   Scenario: One morning, Dorothy checked her mail and noticed a flyer about a new study being conducted by her local health authority to design an online game that people experiencing memory loss can use to help keep their mind active. Ambulatory older adults between the ages of 60–85 years of age were invited to join the study and help researchers design the game. She thought to herself, "What a good idea" and decided to give them a call. Dorothy was enrolled in the study straight away. She was told on the phone by a researcher that she had to go make her way to the downtown Vancouver office. However, on the day, because of the pain in her lower back, she couldn't get to the venue so Dorothy called the researcher who said that she can just call into the meeting instead. Dorothy did as the person suggested and joined the meeting by phone but because she couldn't see anything that was being presented at the meeting she found it difficult to understand what was going on. Because people were talking too fast and using very difficult to understand language, Dorothy got very confused, especially due to her memory issues, and eventually fell asleep on the phone. Dorothy also didn't find their idea to be very interesting because they were trying to make the game like Jeopardy (an iconic knowledge/trivia game in North America) and she really did not like Alex Trebek (the gameshow host). Plus, she thought the game questions were boring and too easy. A few weeks later, Dorothy, along with other participants, decided to quit the study. The research team was very disappointed because it meant that they had to re-recruit and that pushed back the project's completion date.

Solution: In light of Dorothy and some others leaving the study, the research team lead decided to have an emergency debrief with the team members. Looking back, they should have designed the protocol by fully considering the physical, cognitive, and social challenges that may arise when working with older adults who are living with one or more physical and cognitive health conditions to make participating in the study easier. For example, they could have devised a plan to provide transportation for people with mobility problems, avoid jargon during the meeting, and ask participants what their key interests were and if they had any ideas for the game beforehand.

Even though these seem like common-sense solutions, the practical considerations when designing with people with mild cognitive impairment are often overlooked. Working with people with cognitive impairment is not easy especially when they are struggling to learn, maintain, and make sense of new information.

11.6   CO-CREATION IN PRACTICE: CHALLENGES AND LIMITATIONS

Co-creating and working with people experiencing cognitive decline can help to produce valuable contributions not only to the research and design process, but also for the development of products that can have impact in the real world. Past research has found that involving people with dementia as *co-creators* in the research and design process can provide them with "a sense of purpose, value, countering the feelings of powerlessness more usually associated with dementia" (Tanner, 2012, p. 304). However, to ensure the success of co-creation projects, we must recognize and be aware of the main challenges when working with people experiencing cognitive decline. These may include the following.

- Barriers to co-design linked to their cognitive limitations (Hendriks et al., 2014).
    - ◊ For example, difficulties remaining focused in group discussions, making decisions, remembering past discussions and decisions, and increased skepticism and

suspicion of researchers and at times an over-reliance on trusted family members or caregivers in the co-design process.

- Reliability of the information provided by people experiencing cognitive decline especially individuals living with dementia, family members, or caregivers (Meiland et al., 2012).

  ◊ For example, inaccurate recollection of events or situations, over- or underestimating the ability of the person with dementia, or family members' or caregivers' voices dominating the discussion.

- Stressful nature of co-research and co-design (both mentally and physically) for the person with dementia (Grönvall and Kyng, 2013).

  ◊ For example, distress, anxiousness, frustration, frailty, and fatigue experienced by the individual during work sessions.

- Maintaining an equal power dynamic (Lindsay et al., 2012).

  ◊ For example, difficulty balancing power differences between researchers, co-researchers, and the researched, and between people experiencing cognitive decline and their family or caregivers.

- Ensuring psychological well-being of researchers and designers (Hendriks et al., 2014).

  ◊ For example, emotional impact from witnessing situations of distress, pressures, and demands of maintaining good relationships with both people with dementia and their family members and caregivers.

11.7   SUMMARY

People experiencing cognitive decline are often excluded from the research and design process when developing new research and innovative products that could directly benefit them. For some of the reasons identified in this chapter, people with cognitive decline or dementia are known to be a seldom heard group when undertaking research, and when designing new solutions that will enable them to age well at home and in the community. Engaging people with dementia when co-designing new technologies can be challenging, especially regarding difficulties with recruitment, ethics and consenting, time, and resources. Such reasons contribute to the lack of involvement of people with cognitive decline and dementia in the research and design process. Although there is a growing appetite for approaches and methods to improve engagement in technology design, research in this area is limited, and often relies on more general participatory approaches that stem from qualitive research. However, when it comes to the design and evaluation of technology for and with people with cognitive decline or dementia and their social care environment, more considerations are required concerning their individual resources, background, familial situation, and social circumstances. Some people with dementia may also have limited expressive ability, so relatives or carers may need to serve as intermediaries, which, in itself, is very challenging. Therefore, when designing approaches and methods aiming to engage people with dementia, a deep understanding of the symptoms, problems, and requirements of people with dementia as well as the challenges and needs of their carers and other stakeholders, is necessary. Despite the notable challenges of co-creating and working with people experiencing cognitive decline or dementia, using more engaging research and design approaches such as CBPR and KITE has been highlighted as a mechanism for promoting their citizenship and rights—recognizing the value of their experiential knowledge in the co-creation process, and their ability and capacity to contribute in the development of products and services for the greater good.

11.8   FIND OUT MORE

Section III, "Designing together," in the book *Knowledge, Innovation, and Impact in Health: A Guide for the Engaged Researcher*, Sixsmith et al. (2020)—provides guidelines, case studies, and learning activities on co-creation approaches and methods.

OA-INVOLVE. (2020)—Engaging people living with dementia: barriers and facilitators, solutions and tips: http://www.oa-involve-agewell.ca/uploads/1/2/7/2/12729928/dementia_brief_oct_17_2019.pdf.

Ramaswamy, V. and Gouillart, F. (2010). *The Power of Co-Creation: Build It with Them to Boost Growth, Productivity, and Profits.* An easy-to-read account of how co-creation transformed some of the world's biggest companies. New York: Simon & Schuster Free Press.

Astell,   A. J. et al. (2019). Technology and dementia: the future is now, *Dementia and Geriatric Cognitive Disorders*, 47: 131–139. DOI: 10.1159/000497800, includes methods and approaches for working with people with dementia.

CHAPTER 12

Doing Ethical Research with People with Dementia: Challenges and Resolutions

A growing body of literature suggests that many people with dementia can, and do, make important contributions to research (Hedman et al., 2016). However, the notion that people with dementia are too compromised by their illness to contribute (due to deterioration in memory; cognition; behavior; and the performance of everyday and instrumental activities) has limited their inclusion in research (Millett, 2011). However, there are strong, ethical reasons to include people with dementia within research, as they can provide "valuable information on the perspectives and experiences of the people affected by dementia," reduce stigma, support personhood and influence quality of care (Hampson and Morris, 2018, p. 15). Indeed, failing to involve people with dementia can be thought of as disempowering, exclusionary, and discriminatory. This perspective flips the ethical imperative from "Why we should exclude people with dementia" to "Why we should not include them."

Nevertheless, if people with dementia are to be included in the design and production of knowledge, services and technologies, research, and development projects need to satisfy stringent ethical considerations and

> **Box 12.1 The Mental Capacity Act, 2005**
>
> The UK's Mental Capacity Act (UK Public General Acts, 2005) states that people have capacity to make a decision if they can:
>
> ➢ understand the information they are given to make the decision;
>
> ➢ appreciate the consequences of making/not making the decision;
>
> ➢ retain the information;
>
> ➢ use the information to make the decision; and
>
> ➢ communicate their decision.

procedures in order to protect them (as in the case of all research participants) from potential harms that might arise from their involvement. The ethical issues that can arise are particularly challenging when safeguarding the dignity of people with dementia and ensuring the integrity of the research itself, and this is especially the case as each research and development project needs to be seen as a unique situation in which people with dementia are involved (Drageset, 2019).

Alongside the issue of dignity, the involvement of people with dementia in research projects can engender the building of a sense of empowerment if done in ethically sound ways. Despite the need for research and development projects to progress within ethical guidelines and with moral sensitivity, there are no firmly agreed ethical standards that can be applied (West et al., 2017). What is clear is that in the context of research, people with dementia need to be recognized as vulnerable people whose safety needs to be properly safeguarded.

This chapter sets out some of the considerations that need to be thought through when involving people with dementia in research projects so that maximal benefits are achieved and minimal harms occur. Each of the key ideas listed below, while not comprehensive, point to areas for researcher scrutiny. They are meant as a guide to thinking through the main ethical issues faced, rather than being prescriptive of ethical practice:

- mental capacity, informed consent, and assent;

- communication;

- gatekeeping;

- integrity and dignity;

- protection and responsibility; and

- participants or co-researchers.

A final section looks at some of the ethical issues relating to the implementation, adoption and use of AgeTech in the "real world," beyond the laboratory or design studio.

Nokwan007: Shutterstock

12.1   MENTAL CAPACITY, INFORMED CONSENT, AND ASSENT

Obtaining consent to take part in research is an essential element of most research involving people, but is difficult to negotiate when researching with people with dementia as they may or may not have the mental capacity to understand and make a decision to consent. In the example of the UK, this is covered by the *Mental Capacity Act, 2005* (see Box 12.1). It is imperative to make sure people with dementia are not coerced or deceived into participation and are fully informed to volunteer freely (Hampson and Morris, 2018). This means they should have received accessible information in a form that works for that individual. Some will read the information and others will prefer a verbal review of about the aim of the research, their involvement in it, what will be done with their data, how their privacy will be safeguarded, and what the benefits and disadvantages may be both socially and individually and that they can withdraw at any time. Moreover, they should have adequate time to understand this information and agree to it and be enabled to give either written or verbal consent. However, as Hampson and Morris (2018) point out, the capacity to give consent needs to be carefully assessed, as it implies the person's ability to understand the information they have been given, relate it to their own situation, judge the options they have, and understand the consequences of the choices they make. Impairments in cognitive functioning associated with dementia can compromise any or all of these abilities. On the one hand, if informed consent is compromised then the person is vulnerable to exploitation. On the other hand, cognitive limitations should not automatically exclude people with dementia from involvement in research if their protection is assured. Several steps can be taken to maximize the opportunity for people with dementia to give informed consent.

1. Make sure that information given is appropriate to their capacities, and consider connecting this information to their interests, needs, and priorities alongside the most appropriate information medium (written, verbal, pictorial/comic presentation, etc.).

2. To determine cognitive ability to participate or through casual conversation with an experienced researcher use an assessment tool such as the Mini Mental State Examination (Creavin et al., 2016).

3. Establish an ongoing or longitudinal/continuous consent procedure to check if consent is maintained throughout the research. This is very important if the research duration extends over time (e.g., days, weeks, or months) as memory problems may hamper understanding or willingness to continue, the disease fluctuates or deterioration occurs. This withdrawal may be temporary.

4. Negotiated consent can be possible if people with dementia give their assent (such as signals they want to participate), but also defers the informed decision to an advocate (perhaps a family member or carer) (Cobigo et al., 2019).

Finally, while informed consent or assent has been achieved before the research proceeds, it is important to watch out for the person's withdrawal of consent. Easily identified when given verbally, dissent might be discerned through behavioral cues such as silence, lack of cooperation, or agitation and distress and may be temporary reflecting a bad day or time of day. Returning later in the day or another day may give the researcher consent and continuation of the project.

12.2   COMMUNICATION

Communicating with people with dementia in research contexts requires knowledge of the possible communication problems of the participants and the researchers' skills needed to overcome them. Key considerations include:

- repetition;

- lack of understanding of complex sentence structures;

- jargon;

- long silences while the person thinks through their expression;

- forgetfulness;

- incoherence;

- frustration in not being able to express themselves in the way they want to;

- wandering off the topic; and

- switching attention.

However, not all people with dementia will be compromised in these ways and communication abilities can vary during each encounter. It is the role of the researcher to assess communication skills and plan how they will deal with communication difficulties as they arise. Some strategies are: simplifying expressions and ideas if required; avoiding acronyms and abbreviations; signaling informality and relaxation in the conversation; reducing pressure on *performance*; allowing time for the person to recover from any confusion as well as time to think; offering reassurance and acknowledgment of the importance of their contribution; and engaging in small talk. The key element is to be accepting of communication limitations and attentive to variations in the people with dementia's ability to engage with the research and have pre-thought out plans on how to deal with such difficulties.

12.3   GATEKEEPING

Gatekeeping was discussed in Chapter 11, but there are also important ethical issues involved when researching with people with dementia. Dempsey et al. (2016) give good advice in relation to negotiating access to people with dementia through gatekeepers (who can be family, informal or formal gatekeepers), a key aspect of involving them in research. They suggest supplying information appropriate to the gatekeepers about the aims, process, involvements, and benefits and risks of the research. Also, making personal contact is more effective than remote or electronic contact as this helps to develop trust and enable questions to be asked and answered. Acknowledge that gatekeepers may have their own agenda or anxieties about what the research may reveal and work with them to address all anxieties. If they decide not to enable the research, this decision should be respected, even if this means that some people with dementia will then miss the opportunity to have their voices heard. Where possible, the researcher should have a one-on-one conversation with the people with dementia before accepting withdrawal. Sometimes, it is essential, preferable, or an expressed condition or wish of participants that gatekeepers are present during data collection. This could either enhance or compromise the quality of the data. If the person or their gatekeeper asks for this, then it is up to the researcher to decide if this is appropriate in the context of the research topic. Continuing with the research or walking away when such requests are made should be made clear in the ethics section of research's funding application. Finally, it is important, once the research has concluded, that feedback for the findings is made available in accessible ways to gatekeepers as well as the people with dementia themselves.

12.4   EMPATHY AND UNDERSTANDING

In order to ensure that people with dementia are well considered, and that the research is conducted with integrity and is mindful of their dignity, it is important to ensure that researchers have a good knowledge of dementia as a disease to better plan for the complex clinical, mental, and behavioral issues they could meet during data collection. Incomplete knowledge can hamper this and may lead to research that is both confusing for people with dementia and unproductive for the researcher. For example, knowing when the person takes their medication (and thus less likely to be agitated), when they are usually most fatigued (thereby less able to concentrate), when relatives can be present to provide support, and where they feel most comfortable and with the least amount of distractions (e.g., noise) sets up the conditions for participants to be in a positive mindset to engage effectively (Drageset, 2019). Finally, developing trust with people with dementia who are research participants, and their relatives, can help to set-up a constructive research situation. However, it must be considered that

Box 12.2 "Hear it from the horse's mouth"

Jim Mann is a long-time advocate for increasing the role and voice of people with dementia in research. Jim is on the Research Management Committee at AGE-WELL NCE and is on the Board of SFU's Science and Technology for Aging research Institute. Jim is also someone who has been living with dementia. You can listen to Jim presenting about how people with dementia can be meaningfully engaged in research and how their role has been limited by the stereotype of them being "incapable and incompetent." In the video, https://www.youtube.com/watch?v=W8L71l0fnJo, he highlights how Research Ethics Boards are often overly protective, leading to people with dementia being excluded from taking part in research that is directly addressing their own problems, needs, and aspirations.

trust can be a double-edged sword if it promotes inappropriate disclosures or creates unachievable hopes and expectations of the research. Being fully transparent about the potentialities of the research and communicating this in language appropriate to each individual participant is essential.

12.5   PROTECTION AND RESPONSIBILITY

Researching with people with dementia can cause them distress, for example, in their remembering past experiences and relationships in their life that they no longer can enjoy (or indeed they never enjoyed in the first place!), realizing they have dementia when they had been unaware or had forgotten, facing their future with dementia, or the impact their dementia has upon their family. Being aware of the potential for upset, and how this might manifest within the research itself is important. This requires knowledge of dementia, knowledge of the person with dementia and their situation, as well as knowledge of different strategies that can be brought into play to alleviate the situation. Dempsey et al. (2016) suggest the creation of a distress protocol that outlines the conditions: (1) under which the research will be terminated; (2) when researchers need to intervene to give the person time to recover; (3) when the distressing issue will be discussed or the topic changed; (4) when the interview will be need to be rearranged; and (5) when the person will be signposted to support groups who can provide help. In this last case, a list of contact information for support organizations should be complied in readiness for this eventuality. It is important, when distressing situations occur, that the researcher has developed clear guidance on when they should step back from the research and become a concerned citizen. Some of these issues are illustrated in the persona and situation below.

Tom Wang: Shutterstock

Exiting the research situation constitutes a further ethical consideration. Since research relationships with people with dementia can be complex, especially when repeated contact is made (such as what can happen in clinical trials or in qualitative data collection), trust and dependency can develop and friendships formed. In these circumstances, researchers are encouraged to produce an exit protocol to guide them in exiting the research, making clear that the research will end. This might include: when and how to inform the person that the research is terminating and that contact with the researcher is most likely to cease; offering other opportunities to engage in research (if this is possible and desired from the perspective of the participant); and even continuing personal contact outside of the research context.

12.6   PERSONA AND SCENARIO—EDITH

Persona: During her working life, Edith was a General Practitioner in a busy urban practice. She was diagnosed with dementia three years ago and has been living successfully on her own since then. Recently, she has been getting more forgetful, especially about taking her medications and also has had difficulty in finding her way to the shops and to visit friends. This has left her feeling despondent and lonely.

Scenario: Edith learned about an opportunity to be involved in a research project that aimed at developing a technology to help people to connect with their family and friends. This seemed to be a good way to improve her own situation and that of others and so she signed up to take part. When the researcher called on her to gain her informed consent, Edith was happy to comply and agreed to be interviewed about her own situation. One week later, the researcher returned to conduct the interview and while Edith knew she had consented to an interview, she struggled to remember what the research was all about. As the interview progressed, Edith became more agitated as she talked about her feelings of despondence and loneliness. She finally broke down in tears when the researcher asked her about her experience of living with dementia. She had forgotten her diagnosis, and was horrified and shocked when faced with the knowledge that she had dementia.

Solution: A number of key opportunities were missed which might have avoided this upsetting situation. First, the aim of the research and why Edith had been a chosen volunteer could have been clarified at the beginning of the interview. Second, the researcher should have had sufficient knowledge of dementia to identify Edith's increasing agitation. The researcher should have checked with Edith if she was comfortable enough to continue or if she would prefer to stop, using the distress protocol prepared prior to interviews taking place. Third, a list of support organizations could have provided Edith with some contacts to give her support with her despondency and loneliness. Last, the researcher need not have reminded Edith of her diagnosis and simply could have asked her about how she was managing in her everyday life instead.

12.7   VOICE AS PARTICIPANTS, CO-RESEARCHERS, OR PARTNERS

Just as researchers have recognized the value of people with dementia as research participants and have expanded research designs to include their voice, situations, and needs, including people with dementia in project teams as co-researchers or partners is becoming recognized. Not only does this enable research to benefit from their lived experience, it can also be a positive force in supporting their sense of empowerment. The more integral inclusion of people with dementia as advisors and decision-makers in research can help shape research to be highly relevant to other older adults (OA-INVOLVE, 2019). Much of this chapter has concentrated on ethical issues when including people with dementia as participants, however many of the issues discussed also apply to their involvement as co-researchers or research partners. When this is the case, researchers need to give consideration to issues of power, and the influence of power inequalities in all study forums, including team meetings, knowledge translation events, and dissemination opportunities.

12.8   ETHICS AND COGNITIVE HEALTH

The chapter deals primarily with people with dementia as adults within the research context. However, the wider issue of cognitive health includes healthy cognitively intact people, MCI, and early-stage dementia. Having presented the key ideas that underpin rigorous ethical consideration when researching with people with dementia, one further concern is worth discussing. Pierce (2010) pointed out that some research on dementia may be required with people who are currently healthy, but who are known to be at risk of dementia, perhaps because of family history. Targeting research before symptoms manifest (pre diagnosis) can reveal new insights into dementia treatments. However, research with *healthy* people needs to consider the possible complex vulnerabilities that can arise if *healthy* people develop enhanced sensitivities to their own risk of irreversible dementia, perhaps increasing fear and threat of an immensely difficult and uncertain future. Consequently, as Pierce (2010) warns, for those at risk of dementia, but who may not actually develop dementia, researchers need to concentrate on maintaining their safety. Here, scrupulous attention needs to be paid to: (1) ensuring the benefits of the research outweigh to potential distress the research may cause; (2) informing participants of the potential, sometimes unintended, impacts of taking part in the research; and (3) providing any support emotional or informational support as needed.

12.9   TECHNOLOGY IN THE REAL WORLD

This chapter has primarily focused on ethical aspects relating to research involving people with dementia. A number of other ethical issues arise when considering research that develops technologies aimed at improving their lives. This extends ethical considerations to the post research situation. How technology is implemented, sold, and used beyond the laboratory is an important issue within AgeTech more generally, and people with dementia specifically (Sixsmith, 2006). For example, ethical issues should be considered concerning: (1) the potential misuse of technology, as well as areas where misuse may happen, e.g., use of surveillance technology in the home reducing achievement of privacy; (2) data protection and maintenance of confidentiality; (3) alignment of technology with national regulations; and (4) use of technology in everyday life and care contexts. Sixsmith (2006) highlighted six ethical dilemmas related to how new technologies might be used in care practice.

1. Does the technology afford a wider range of services or just constrain choice in other ways?

2. Does it mean better services or an inevitable loss of privacy and dignity?

3. Would services become more responsive to the individual or become more impersonal through the use of technology?

4. Does a new product meet the real needs of consumers, or just reflect marketing strategies aimed at exploiting vulnerable people?

5. Would a technology be empowering or contribute to increasing the dependency of the person?

6. Would any financial savings made through the technology be reinvested in products and services or would it be a cost-cutting exercise?

One issue that is important when considering public services (health, transport, etc.) is where access and use is seen as a right of everyone. For example, a device may be too complicated for some people with dementia to use, or a useful product may be too expensive for many people to afford, even if there is a clear need. These kinds of issues contribute to the *digital divide* (Fang et al., 2019) where certain groups of individuals, such as people with dementia, are disempowered and sometimes excluded from the benefits of accessing or using a technology.

It is not feasible to explore these kinds of real-world ethical issues in depth in this chapter and the reader is encouraged to explore these further in the supporting resources that follow. Researchers working in the AgeTech sector, especially when researching with people with dementia, should think about ethics not a set of rules but as an ongoing dynamic process of negotiation throughout the research process and beyond.

Box 12.3 Ethical by Design

Longeway (2019) argues that design should go beyond the basic ethical guidelines that are set by regulatory bodies. Their ethical by design manifesto aims to support developers, providers, and users in the collaborative process of inherently and explicitly including ethics into product and service design. Researchers and developers should not see ethics as someone else's job. The manifesto proposes a set of principles acting as signposts for developers to consider, discuss, and support at all stages of product development, for example:

➤ providing enough information for people to make informed decisions;

➤ respecting people's right to choose how they engage with the product or service;

➤ complementing differing needs, abilities, viewpoints, and morals; and.

➤ supporting shared decision-making and feedback.

12.10   CONCLUSION

Researching with people with dementia is essential to ensure their voice is heard and their experiences, perspectives and needs inform research findings and developments in the real world.

• Gaining their informed consent to participate is essential.

• Researchers need to have knowledge of dementia and be flexible in terms of their communications skills and abilities to cope with any distress that might occur.

• Remember that people with dementia have rights, but as vulnerable adults, their safeguarding is paramount and takes precedence over the research. Ensure all research considers issues of integrity and dignity.

• Be prepared to deal with difficulties that might occur during research through production of protocols such as a distress protocol and an exit protocol.

• Consider that people with dementia might be capable of acting as co-researchers and partners in research as well as participants.

• How technology may be used in real-world situations needs to be considered as ethical issues in research projects

12.11   FIND OUT MORE

For further ideas and support relating to ethical dilemmas and practice when researching with people with dementia, see the following.

Sixsmith, A., Sixsmith, J., Fang, M. L., and Mihailidis, A. (Eds.) (2020). *The Book Knowledge, Innovation, and Impact: A Guide for the Engaged Health Researcher*. Springer; contains chapters, case studies, and learning activities related to ethical aspects of technology research.

More on the issue of gaining consent can be found in:

Higgins P. (2013). Involving people with dementia in research. *Nursing Times*, 109(28): 20–23: https://tinyurl.com/y3exah8s.

The UK's National Institute for Health Research (NIHR)—INVOLVE website contains useful information on the involvement of people with dementia in research studies: https://www.invo.org.uk/?s=dementia.

For further reading on (1) operationalizing ethical challenges and (2) proxy consent read:

West, E., Stuckelberger, A., Pautex, S., Staaks, J., and Gysels, M. (2017). Operationalising ethical challenges in dementia research—a systematic review of current evidence. *Age and Ageing*, 46(4): 678–687: DOI: 10.1093/ageing/afw250.

Dubois, M. F., Bravo, G., Graham, J., Wildeman, S., Cohen, C., Painter, K., and Bellemare, S. (2011). Comfort with proxy consent to research involving decisionally impaired older adults: do type of proxy and risk-benefit profile matter? *International Psychogeriatrics*, 23(9): 1479–1488. DOI: 10.1017/S1041610211000433.

The Ethical by Design Manifesto can be found here—Mulvenna, M., Boger, J., and Bond, R. (2017). Ethical by design: A manifesto. In *Proceedings of the European Conference on Cognitive Ergonomics 2017 (ECCE 2017)*. New York: pp. 51–54. Association for Computing Machinery (ACM). DOI: 10.1145/3121283.3121300.

CHAPTER 13

Policy, Technology, and Cognitive Health

Cognitive health has emerged as one of the major health, social, and economic challenges of the 21st century and is increasingly the focus of government health and social policy. A key concern is how health care systems will be able to cope with increasing numbers of people with dementia in coming years.

- "The imminent growth in the number of people living with cognitive impairment will place significantly greater demands on our systems of care" (Centers for Disease Control and Prevention, n.d.).

- "People with cognitive impairment report more than three times as many hospital stays as individuals who are hospitalized for some other condition" (Centers for Disease Control and Prevention, n.d.).

- "In 2007…Alzheimer's disease was ranked as the 7th leading cause of death among American adults aged 18 and older. Since then, it has surpassed diabetes to become the 6th leading cause of death among adults, and continues to be the 5th leading cause of death for those aged 65 and older" (Centers for Disease Control and Prevention, 2011).

- "Dementia is one of the main causes of disability later in life, ahead of cancer, cardiovascular disease and stroke" (Chambers, Bancej, and McDowell, 2016).

In a country such as Canada, for example, it is projected that by 2031 the total annual health care costs for Canadians with dementia will have doubled those from two decades earlier, from $8.3 billion to $16.6 billion (Government of Canada, 2017). While dementia represents the most critical challenge, policy initiatives are also needed to support healthy cognitive aging and people with MCI and early-stage dementia at home.

13.1 COGNITIVE HEALTH IS A WORLDWIDE CHALLENGE

Cognitive health is a challenge that is being faced not only by the richer, economically advanced countries in North America, Europe, and Asia, but increasingly in the middle- and lower-income countries across the world. According to the WHO, approximately 50 million people worldwide live with dementia (WHO, 2020). A major push for policy change surrounding cognitive health, specifically concerning dementia, has come from the WHO and Alzheimer's Disease International. In May 2017, a global action plan on the public health response to dementia 2017–2025 (Alzheimer's Disease International publication team, 2018) was adopted that includes policy recommendations and guidelines for countries to develop tailored response plans to dementia. Recommendations include (Alzheimer's Disease International publication team, 2018) the following.

- Policy changes for a cognitive friendly society can be accomplished through partnerships and commitment among various organizations and individuals.

- National dementia plans should include meaningful engagement from all stakeholders including people with dementia and their caregivers, to give plans a multidisciplinary and holistic approach.

- Countries should implement at least one public awareness campaign to facilitate a dementia inclusive society and raise awareness for cognitive health.

- Priorities of these campaigns should involve bringing awareness to the challenges individuals with cognitive health declines face and the stigma surrounding changes in cognitive health.

- Both the physical and social environments of people with cognitive health issues should aim to be inclusive and supportive of the needs of individuals with cognitive health declines.

- These goals and policy changes for a dementia-friendly society can be accomplished through partnerships and commitment between various organizations and individuals.

13.2   SOCIAL PRIORITIES

Continued advocacy and dissemination of information concerning cognitive health from global organizations such as Alzheimer's societies worldwide, the WHO, and the United Nations is beneficial in bringing about social change and quality of life impacts. The social impact of better cognitive health is far-reaching. Through supporting the cognitive health of individuals they are more likely to be able to participate within their communities and continue to pursue work, and vocational and leisure activities longer. This support enables individuals to stay connected and to retain a sense of purpose in their lives. Supporting cognitive health is also an opportunity to destigmatize the narrative around declining cognitive health and to better demonstrate how cognitive health varies. Better education about cognitive health will help to promote healthy cognitive aging and encourage people to detect cognitive decline and seek support earlier. Further, education and awareness of cognitive health needs will lead to more accessible resources and programs for both individuals and communities to support cognitive health. Overall, raising awareness about cognitive health can allow people impacted by cognitive health changes to be more in control of their own health and to reduce the stigma associated with dementia.

13.3   ECONOMIC PRIORITIES

The economic impact of dementia has already been mentioned, as has providing cost-effective solutions to meet the needs of the increasing numbers of people with dementia. This has been driving the growth of research and innovation in AgeTech and technology-based solutions to support people with dementia, which have been discussed earlier in this book. The economic impacts of dementia and cognitive decline are wide-ranging. The maintenance of occupational productivity from both informal caregivers and the individual experiencing cognitive health declines. Many individuals who experience cognitive decline are less likely to be able to participate as actively in the economic market as they did previously, whether this was through occupational work, volunteering, travel and leisure, or through spending habits. Similarly, as cognitive decline progresses many family members need to take unpaid time from work or leave their jobs to provide care for their loved ones (Family Caregiver Alliance, 2018). Both of these scenarios have an impact on the economy. By supporting cognitive health individuals can continue to participate in economic development for a longer period of time. In a similar line of thought, education for business and employees on how to serve customers with cognitive health declines can broaden the range of services and goods individuals may seek out due to acceptance and credibility.

13.4   NATIONAL PLANS FOR DEMENTIA AND COGNITIVE HEALTH

The last decade has seen many countries develop national strategies to meet the challenge of dementia (Alzheimer's Disease International, n.d.). Most of these are wealthier countries, such as the U.S., Australia, United Kingdom and other European countries, all of which have a long history of developed policy in relation toward old age and population aging. For example, Alzheimer Europe provides a comparison of 21 countries that have developed national dementia strategies (Alzheimer Europe, 2013).

Box 13.1 Canadian Dementia Strategy

The implementation and development of some of these policy priorities have been taken up in Canada's draft for its national Dementia Strategy (Public Health Agency of Canada, 2018). The discussions and collaboration of involved stakeholders needs to continue in order to focus on policies that will benefit Canada as a whole when implementing policies for cognitive health. Continued partnerships such as the Canadian Alzheimer's Disease and Dementia Partnership (Canadian Alzheimer's Society, 2017) are critical to spreading awareness and applying pressure for policy changes. For example, this organization concentrates on mandate changes required to target the issue of Alzheimer's Disease and dementia in Canada, while focusing on three mandate elements: research, prevention, and living well with dementia. It discusses these elements through their key initiatives, how they can measure their progress through targets, and how this will benefit Canadians. Continued partnerships among reputable organizations, as well as the community, are critical for moving toward policy change.

13.5   SUSTAINABLE DEVELOPMENT GOALS

The United Nations has set out an ambitious agenda for global Sustainable Development Goals (SDGs) for 2030. The SDGs address the key challenges facing all countries, such as poverty, inequality, and climate change. The aim is to achieve them by 2030 in order to "leave no one behind" (United Nations Sustainable Goals, n.d.). Aging is seen as an important aspect of the SDG agenda: "Preparing for an aging population is vital to the achievement of the integrated 2030 Agenda, with aging cutting across the goals on poverty eradication, good health, gender equality, economic growth and decent work, reduced inequalities and sustainable cities. Therefore, while it is essential...to go beyond treating older persons as a vulnerable group. Older persons must be recognized as active agents of societal development in order to achieve truly transformative, inclusive, and sustainable development outcomes" (United Nations. Department of Economic and Social Affairs Ageing, 2017).

The SDG agenda argues for a very different approach to aging. Rather than focus on the problems of ill-health and dependency, the approach is about rights, and how all older adults, even those with dementia, can and should be enabled to be participants in society. AgeTech also needs to align with this approach, to design and implement technology-based services, and to support this agenda of social participation.

13.6   POLICY AND THE ROLE OF TECHNOLOGY

Until relatively recently, dementia and cognitive health were not a major part of the AgeTech research and innovation agenda (Sixsmith, 2006). However, over the last decade, technology has become increasingly recognized as a facilitator for change and innovation, improving the range and quality of services and increasing the capacity to meet the needs of growing numbers of people with dementia. In the next sections in this chapter, there are highlights of some of the global and national policy priorities that have emerged in recent years.

13.7   HEALTH SERVICE ORGANIZATION AND DELIVERY

Providing services for people with dementia and maintaining cognitive health in aging is a complex topic. This complexity in maintaining cognitive health raises some questions and challenges that need to be addressed.

- Can technology address the diversity and variability in cognitive health changes in aging through personalized health care delivery?

- How technology can be used to deliver support to underserved and marginalized groups.

- Can technology be used in areas of health care, such as community mental health, that have been underfunded and ineffective?

- Can technology support cognitive health globally, especially for those in rural and remote areas?

Matej Kastelic: Shutterstock

13.8   FOCUS ON PREVENTION

If the personal and social stresses of cognitive impairment and dementia are to be reduced, then public health policy should focus on risk reduction through dementia prevention and health promotion. Promoting awareness and initiatives to reduce the risks of dementia would see policy that increases physical activity, decreases smoking and over consumption of alcohol, and targets key risk factors related to dementia. The production and promotion of educational material concerning cognitive health for both individuals and caregivers could help reduce the stigma that currently prevents people from seeking aid. Questions from the prevention standpoint include the following.

- How can risk information be distributed effectively to promote early cognitive health practices?

- Can screening tools be developed to identify those at risk of cognitive health declines?

- How can both individuals and caregivers be educated to recognize warning signs of cognitive health declines?

13.9 SUPPORTING PEOPLE WITH DEMENTIA

Care for cognitive health is complex, made even more so by individual variations. The care for people experiencing cognitive decline should reflect this with personalized health care recommendations.

- Can technology help dementia and caregivers to access information and supports as early as possible?

- What can be done to best stabilize or maintain the cognitive health of those experiencing declines?

- How can the health and social care services be integrated between the individual, formal, and informal sectors?

- How can we account for culturally different views of cognitive health in best care practice?

- How can we remove the stigma that comes with dementia?

13.9.1 SUPPORTING CAREGIVERS

Dementia is an age-related health condition that places huge demands on family caregivers, especially when they provide unpaid care. Indeed, formal health care services only provide a fraction of the care needed; previous studies have suggested that two-thirds of the cash and in-kind costs fall on families (Alzheimer's Society of Canada, 2016). This focuses attention on questions around technological solutions regarding financial and social support.

- How can caregivers be helped in financial and employment terms?

- How can technology be designed to strengthen informal support networks?

- What is the best way for technology to provide education and resources to facilitate high quality informal care?

13.10 WIDER ECONOMIC CONSIDERATIONS

The economic impact of technologies for maintaining and supporting cognitive health has wide-reaching impacts, from an individual level to a national one.

- How can AgeTech be ensured to help older adults with MCI and age-related cognitive changes and informal caregivers to continue in their working careers for longer?

- How can technology increase productivity and well-being in the workplace?

- Can AgeTech provide more economically sustainable models of dementia care, by increasing community-based support?

13.11   ETHICAL ISSUES

It is important that technological interventions are developed and implemented in an ethically appropriate way. Key issues include (ethical issues are addressed in more detail in a separate chapter):

- how to ensure human connection remains intact when designing technologies;

- how to make technology age-accessible for a generation that did not grow up with it;

- how to ensure technology to support cognitive health is usable at any level of cognitive health; and

- how to ensure that technology for cognitive health is affordable and accessible to those who need it.

13.12   SCIENCE POLICY AND RESEARCH

Dementia and cognitive health has become a major focus of research globally in recent years. While much of this effort has focused on understanding the brain science, i.e., the causes and cures, behind dementia and developing pharmacological treatments, limited progress has been made on either of these two fronts and there is limited prospect for a "cure" for dementia in the short and medium terms. This highlights the

GaudiLab: Shutterstock

need for research that is aimed at informing policy, improving service delivery, enhancing care practice, providing non-pharmacological interventions, and developing novel supports for people dementia and caregivers. AgeTech has, in its own right, become a major focus of investment in research and development (Sixsmith et al. 2017). As has been seen with AGE-WELL NCE and initiatives in the European Union, much of this investment has been on innovation-focused research aimed at understanding real-world problems, developing solutions, and accelerating deployment and uptake.

Box 13.2 Global Dementia Observatory

With the scientific advances of technology the ability to collect and share information concerning cognitive health and over a larger set of individuals will increase. This is currently already unfolding with the Global Dementia Observatory (GDO) organized by the WHO (2017c). The GDO uses information that is collected globally from the various countries that have implemented Global Dementia Action Plans. This information can be used to strengthen or inform policies and plan for future research for cognitive health care. This data could be further used to potentially impact the quality of diagnosis and prognosis for cognitive health declines. In the context of cognitive health and artificial intelligence (AI) technology, large-scale data could be used for predicting which combination of risk factors may lead to dementia or how to best manage risk factors to prevent declines. The impact of focusing on cognitive health on the scientific community has many far-reaching implications that need to be further explored.

13.13   COST EFFECTIVENESS AND IMPACT

One of the biggest selling points of technology-based interventions is that they have the potential to help to reduce health care costs in the face of increasing numbers of people with dementia. Supporting people in the community has long been seen as a lower-cost option compared to expensive residential and hospital care (e.g., Sixsmith and Sixsmith, 2008). Costs may be reduced by prolonging the time people can remain living in the community (i.e., reducing demand) and by providing lower-cost alternatives. But what is the evidence for this? Can an economic argument be made for technology-based interventions generally and for those that are specifically targeted at older adults with dementia? What is the evidence for impact on quality of services? What evidence is there for health and quality of life outcomes? Policy to support technology adoption needs this evidence if it is to be effective. This will be addressed more specifically in the next chapter.

13.14   POLICY NEXT STEPS

Having supportive policies at the national and local levels is essential if useful technologies are to be scaled-up, implemented, and adopted by the people and organizations that will benefit from them. This remains a significant challenge in the AgeTech sector generally and especially so with people with dementia and cognitive decline. For example, the national dementia strategies that are being put in place in many parts of the world need to have an explicit AgeTech component given the importance of information and communication technologies in virtually every aspect of life in the 21st century.

13.15   FIND OUT MORE

There are numerous useful sources to find out more about dementia and cognitive health and the policy challenges and opportunities for providing better health and social care as well as supports to those in need.

Alzheimer's Disease International. Dementia plans—This website discusses the plans of European countries as well as territories regarding dementia: https://www.alz.co.uk/dementia-plans.

Alzheimer Europe—*2018: Comparison of national dementia strategies in Europe*: https://www.alzheimer-europe.org/Policy-in-Practice2/Country-comparisons/2018-Comparsion-of-National-Dementia-Strategies.

Alzheimer's Disease International publication team—From plan to impact: Progress toward targets of the global action plan on dementia (2018). Report: https://www.alz.co.uk/adi/pdf/from-plan-to-impact-2018.pdf?2.

Prevalence and Monetary Costs of Dementia in Canada (Alzheimer's Society of Canada, 2016)—Provides an overview of dementia and discusses how common dementia is in Canadian society and why that is. It also reviews the financial burden that dementia places on not just the ones living with dementia, but also their caregivers and families. https://www.canada.ca/content/dam/phac-aspc/migration/phac-aspc/publicat/hpcdp-pspmc/36-10/assets/pdf/ar04-eng.pdf.

The healthy brain initiative: The public health road map for state and national partnerships, 2013–2018. Alzheimer's Association and Centers for Disease Control and Prevention (2013)—Discusses how public health agencies can improve cognitive health for the individual affected, the community, and care partners. The report overviews cognitive health in detail and discusses action items to improve cognitive health. https://www.cdc.gov/aging/pdf/2013-healthy-brain-initiative.pdf.

CHAPTER 14

Demonstrating Impact—Is Technology Effective in Supporting People with Dementia?

While technological innovation tends to focus on engineering and information technology (IT) development, a key area for research is in the evaluation and demonstration that new technologies and solutions are beneficial, effective, and provide value for money. Of particular interest to health care providers are pilot studies, field trials, and evaluations to see whether the intervention or innovation has: (1) clinical and quality of life benefits; and (2) cost savings over alternative treatments. These also benefit a company or research team by providing real-world feedback that allows them to bring a new device or technology to market, as well as providing marketing information. However, there is a lack of real-world evidence to support the adoption of AgeTech generally and for people with dementia specifically.

14.1 LACK OF EVIDENCE TO SUPPORT TECHNOLOGY ADOPTION

A Cochrane Review carried out by Martin et al. (2008) looked at the evidence base for technology adoption within health and social care. The review produced a significant volume of literature on the use of smart technologies within health care. However, there were no studies showing smart technologies' effectiveness and the (potential) outcomes of smart technologies to support people in their homes (see Liu et al., 2016). This highlights the need for better research. An early study by Wootton and Hebert (2001) found no important health outcomes in a randomized control trial of a technology intervention. This lead to the conclusion that telehealth does not seem to be a cost-effective addition to existing standards of care. Grustam (2014) noted that studies of technology-based interventions fail to present economic evaluations that compare costs and effects of a technology-based intervention with a comparator, while data on investment costs are typically lacking. Those studies that assessed costs and outcomes did indicate some limited benefits, but the methodological quality of the studies was typically low. More recently, Krick et al. (2019) noted the continuing lack of evidence and argued for high-quality evaluations of existing technologies in real-life settings rather than systematic reviews with low-quality studies.

BOX 14.1 United Kingdom Impact Analysis of Telehealth and Telecare

One country that has been very active in looking at the evidence base for adopting technology is the UK. The UK Department of Health's Whole System Demonstrator (WSD) Pilot Programme, was a randomized control trial undertaken in England to evaluate the clinical effectiveness and cost effectiveness of telecare and telehealth (Henderson et al., 2013). The objective of the WSD Programme was to provide evidence to support the commissioning and delivery of telecare and telehealth services through generating statistically significant findings that would provide more robust evidence than the case studies and small-scale trials that have been conducted thus far. Despite some initial promising results, the WSD program failed to provide the definitive clinical and cost-effectiveness evidence to support widespread adoption of technology based services. This highlights the challenges in implementing large-scale trials of AgeTech and the need to continue to build an evidence base.

14.2   IMPACT OF TECHNOLOGY ON DEMENTIA CARE AND SUPPORT

While focusing on dementia care, researchers argue that while technology is promoted by the Department of Health in England as a means of supporting older adults at home, these claims are based on limited qualitative evidence or small-scale quantitative studies (Leroi et al., 2013). Leroi et al. (2013) outline the kinds of protocols needed for randomized control trials to answer questions about the efficacy and cost-effectiveness of technology to affirm that it may positively influence quality of life (QoL). Even the few limited studies that have been carried out, such as Nijhoff et al. (2013) and Wray et al. (2010), have not been able to provide clear evidence to support large-scale adoption of technology-based solutions.

The need for a stronger evidence base is crucial. However, there are a number of challenges. First, designing trials that provide conclusive evidence of the benefits and cost-effectiveness of technology is extremely challenging, as was shown by the WSD in the UK. Indeed, one could argue that large randomized trials are not the gold standard they are meant to be. This is especially true when technology-based services and interventions are complex or are aimed at long-term preventative outcomes. A second problem is that a focus on financial benefits may be missing the point. Technologies that address the health, independence, and QoL of older adults with dementia and their caregivers may not have direct economic outcomes. Perhaps the impetus for adopting AgeTech solutions should be more clearly focused on QoL benefits? A third problem facing the AgeTech sector is that even if financial benefits can be demonstrated, investment does not necessarily follow. For example, investment in cost-effective community-based services would require de-investment in other areas, such as hospital care. Finally, technology is always in flux, even if research in the field shows benefits. Thus, the AgeTech sector is likely to always be lagging behind cutting-edge technology.

14.3   CONCLUSION

One of the biggest challenges in the AgeTech sector is the lack of strong evidence to support the adoption of technologies. This lack of evidence for the benefits and cost effectiveness of new solutions is undermining efforts to innovate within the AgeTech sector. If health providers and funders do not see a compelling financial case for adopting technologies, then there is little motivation to do so. Despite this, it is hard to imagine that future innovation within the care sector, generally, will not have a strong technology basis. A major direction for research in the AgeTech has to be an increased focus on demonstrating its impact.

CHAPTER 15

Commercialization and Knowledge Mobilization

A key idea in this book is that commercialization and knowledge mobilization should be an integral part of a project from the very start. This may sound premature, but for a relatively small investment of time and resources, there may be significant payback later on by helping to avoid expensive mistakes and developing a strong vision for how a research idea can move "from the laboratory to the real world." With increasing numbers of people living with dementia, who want to stay living at home and to enjoy a good quality of life, this is an expanding market for AgeTech products and services. We will not be addressing issues such as intellectual property (IP) management or different pathways to commercialization, such as licensing or start-ups. What we will do is to provide some basic ideas and guidance about how to better:

- understand the customer;

- understand the market; and

- develop a simple early stage business model.

Many of the ideas are applicable generally to commercialization and knowledge mobilization. However, this chapter will pick out key issues that relate specifically to cognitive health and dementia. To start with, this discussion will look at a persona and scenario of Ryan to illustrate the kinds of situations that might push us to think more "commercially."

15.1 PERSONA AND SCENARIO—RYAN

Persona: Ryan is an 80-year-old widower who lives in his own small house in Edinburgh, Scotland. Ryan doesn't have any children and since his wife died three years ago, he has lived a very isolated existence with few friends. But he has always been a bit of a loner and rarely feels lonely. Over the last year he found that he was forgetting things and was even getting lost when he was walking to the shops. Being a practical type of person, he went to his family doctor to discuss his problems. The doctor did some tests and referred him to the local memory clinic. He was told that he had early signs of Alzheimer's disease. This worried him a lot because he was unsure how he would be able to manage on his own if his memory got any worse.

Kristina Kokhanova: Shutterstock

Scenario: Because he lived alone, the memory clinic suggested that he took part in a trial of a new system that was being trialed by one of the local universities. He was told it would "keep an eye" on him and connect him to local community services if he had a problem. Ryan thought this was a great idea and agreed to sign up. A few days later, one of the research engineers came to Ryan's house to explain things to him and ask him to sign a consent form. Unfortunately, Ryan became quite upset when the researcher said that cameras would be installed in various parts of the house and told the researcher that he "didn't want people spying on him." The researcher explained that the "data from the vision sensors would be analyzed by machine learning algorithms" and that no one would actually be looking at the actual images from inside his house. However, Ryan wasn't convinced and commented "They're still cameras aren't they?" Ryan decided he didn't want to participate in the trial. The research team received a lot of negative feedback from potential users during the recruitment process and decided that they needed to go back to the drawing board.

Solution: In hindsight the team realized that while the basic concept was good, but the use of cameras would not be acceptable to potential future customers. They decided they needed to use different sensor technology. They also brought in someone from their design school who said they could probably have saved time, effort and expense by thinking about what their system would look like as a real-world product from the perspective of the customer/user. A basic first step that the team could have done was to create a storyboard using images and simple words to present their ideas to potential customers and get their feedback. The storyboard would help to show and explain how the system would look and how it would work. They also realized that they could have done a better job explaining or "selling" their idea to people like Ryan and provided their research students with workshop on communication skills.

15.2   SOME BASIC IDEAS

The example of Ryan highlights the need to see him as a potential customer and to see the system as a potential product. Of course, researchers need to have the scientific agenda at the forefront of their minds. For example, there may be significant scientific and engineering challenges that they need to solve. However, it is hard to imagine that anyone working in the AgeTech sector would not wish to see their technology being used and contributing to improving the lives of people with dementia.

Product: Researchers and academics might not see their work as creating "products." However, even academic articles and books can be seen in these terms—beyond the creation of the content, there is a complex supply chain of organizations, people, resources, and activities that move the product (article) from supplier (researcher) to customer (reader). Even if this is done in-house (for example if someone publishes a newsletter or blog), it is still something that needs the connection between supplier and customer. AGE-WELL NCE talks about three types of "products."

- **Technology products**—Systems and devices aimed at directly supporting the people with dementia and caregivers.

- **Service products**—These are the delivery models and mechanisms that will allow new technologies and solutions to actually be provided to the user or patient.

- **Policy and practice**—These are knowledge products, such as information, policy briefs, guidelines, standards and regulations, models of good practice, as well as health-related information for the public.

For any product to be useful, and used, it needs to be delivered in a way that is accessible and understandable to the intended user. The product will need to be in a form that target customers (e.g., caregivers, people with dementia, service providers) can readily access/purchase, adopt, and use in their everyday lives and businesses.

Commercialization: "Commercialization" is often seen as a "dirty" word in academic circles, but if a new technology is ever going to get into the hands of people who might benefit from it, then it will have to be commercialized. Commercialization is the process of bringing a new product or service to market and someone will have to "productize," manufacture, package, market, and sell the device, all of which needs financial investment (Battersby and Viswanathan, 2020). This is increasingly important as digital health technologies and assistive devices are part of a growing direct-to-consumer health care market. However, most people working in the academic research fields are totally unprepared for the challenge of product commercialization (Mehta, 2004).

Knowledge mobilization: Knowledge mobilization (KM) is about connecting the right people with the right products and information (Simeonov, Kobayashi, and Grenier, 2017). Terms like KM, knowledge translation, and knowledge exchange are often used interchangeably (Barwick et al., 2014), but they all focus on the flow of knowledge, ideas, and innovation between researchers and stakeholders with the aim of driving change and ultimately social and/or economic impact. Increasingly, the common view of knowledge translation as a one-way linear flow of information from knowledge producers to consumers is seen as too simplistic. KM should be seen as a dynamic and integrated activity as discussed in earlier chapters on co-creation. While we focus on commercialization of technology and service products in this chapter, the same kind of entrepreneurial thinking and action is required with knowledge products. For example, information needs to be "packaged" and disseminated in a manner that can be similar to commercial products.

Technology maturity: This is a way to visualize the way products move from concept to final implementation Advancing innovation requires mechanisms that effectively manage technology development through to its maturation into the market. For example, NASA's Technology Readiness Levels (TRL) (Mankins, 2009) encompasses nine levels of maturity with level one concerning ideas and level nine being technology in its most mature form. A simplified model is the Product Innovation Pathway (Sixsmith et al., 2020) that describes various broad steps: initial ideas, project planning, development, implementation and evaluation, and commercialization. Even at early stages, thinking commercially is important and these ideas should continue to evolve as the project develops.

15.3   VALUE PROPOSITION

A product is something that is useful in real life and if it is useful, it has to reach the people who are going to use it. A value proposition is a brief outline of your ideas about your product. It needs to clearly say what the need or problem is that you are addressing and who the intended customer is. The crucial aspect of this is the "value," the key benefits that the customer will get from buying or using the product.

- No more than 1/2 (one half) a page.

- Clear, straightforward language.

- Needs to capture attention of the consumer.

- All the team should agree and sign-off on this and serves as benchmark for everyone.

- Can be used as the basis for outreach and publicity.

Box 15.1 Research Project Lay Summaries

Researchers are typically expected by funders to provide lay summaries of their projects for consumption by the general public. In a very real sense, a lay summary could be seen as the very first step in commercializing an idea or project. However, most lay summaries are not appropriate for a general readership (Wada et al., 2020). Cutting and pasting from a proposal and writing in academic jargon often makes them unintelligible outside the research community. Wada et al. (2020) created guidelines for how lay summaries can be co-created with stakeholders, following the ideas discussed in earlier chapters. This process should create lay summaries that are more understandable to the intended audience, while at the same time affording an opportunity for stakeholders to shape the ideas and directions of the research.

- It needs to be revised in line with any changes that might happen.

15.4 VALUE PROPOSITION TEMPLATE

- For (**Target customer**)—Insert the specific customer(s) that your product is aiming at. Note that the "user" may NOT be the "customer" who buys or decides.

- Who wants to (**Need**)—Insert the customer's need, problem, or aspiration.

- We developed (**Product**)—state what your product is.

- Our product will (**Features**)—List the key features of the product.

- It will (**Unique feature**)—Say what sets your product apart from the competitors.

15.5 EXAMPLE—VALUE PROPOSITION FOR A NON-INTRUSIVE HOME MONITORING SYSTEM

- For a community dementia care provider.

- Who wants to improve community support for isolated people living with dementia.

- We developed a non-intrusive, in-home monitoring system using simple motion sensors to connect a person with dementia who lives alone to a local call center.

- Our product will:

  ◊ Detect potential alert situations where a caregiver may need to provide timely help.

  ◊ Improve the range, quality, and reach of community supports to people with dementia.

  ◊ Afford greater assurance and security to isolated people with dementia and their families to allow them to remain living at home.

  ◊ Reduce demands and costs on acute and long-term residential care.

- It will monitor people at home without compromising their privacy.

15.6 TECHNIQUES AND METHODS FOR ENGAGING WITH CUSTOMERS

Many of the ideas we discussed about co-creation are relevant here. There are numerous techniques to allow developers to engage with customers at a very early stage, prior to making major investments of time and resources in prototype development. Techniques for helping customers or users to visualize include the following.

- **Persona and scenario**—Short, semi-fictional case studies of target customer/user and their living situations and problems

- **Co-creation workshops**—These are workshops where researchers and stakeholders can work together to articulate problems, identify priorities, and map out high-level solutions.

- **Storyboard**—A simple visual and written/verbal narrative of how a device, system, or service might work in practice.

- **Paper prototype**—Simple cut-out paper models of new devices to show what they might look like to potential users.

- **Mock-up**—A rudimentary model of a device.

- **Wizard-of-Oz technique**—A rudimentary working mock-up of a device that is manipulated behind the scenes by an operator to interact with the person using it.

- **Theater method**—This involves actors performing a scripted enactment of how a new product might operate—this is particularly useful for demonstrating the technology or service to an audience of stakeholders and getting feedback.

All these are generally useful at early stages of the design process, but are particularly useful when working with people with dementia, who may not be able to understand or work with abstract ideas. Personas and scenarios are particularly useful in illustrating and highlighting the human side of what it is to live with dementia. The development of these kinds of relationships can be very important in a project, where stakeholders may be involved on a long-term basis, providing feedback and ideas during the prototyping cycle. Another point is that one should not confuse the customer with the user. The user is the person or organization who will use or operate the product. The customer is the person or organization that is the decision-maker who decides to buy or adopt a product.

15.7   UNDERSTANDING THE MARKET

Understanding the potential customer is important, but so is having a strong awareness of its market sector. Pretty much every AgeTech research proposal will start with statistics about the growing number of older adults in the population, or the numbers of people with dementia, as a justification for research funding. While these high-level figures show general trends, they actually say very little about the actual market for a new product or service. For example, the total number of people with dementia can be referred to as the Total Addressable Market (TAM) for a dementia-related product. While this may look large, there are major segments that impact on who might use your products. For example, the needs of people with dementia who live alone are often very different from someone who lives with a family member. A key concept is the Serviceable Obtainable Market (SOM), i.e., what a company could realistically reach in the short medium and longer terms.

These quantitative data are useful, but any successful innovator or entrepreneur needs to be aware of the micro- and macro-forces, and the drivers that are operating in their sector.

- Where is it coming from?

- What's in there now?

- Where does it seem to be going?

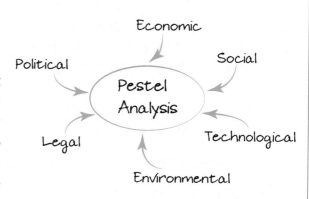

There is no gold standard for doing this kind of work, but we outline three techniques that are useful: *Environmental Scan*; *PESTEL* (Political, Economic, Social, Technological, Environmental and Legal) analysis; and *SWOT* (Strengths, Weaknesses, Opportunities, and Threats) analysis.

Environmental scan—It is important to know what is going on in a sector to look at technologies that address a problem area that are already on the market or in the pipeline. This means having to quickly compile information about the landscape—the way it looks now and the way it might shape up in the future and how a product might fit in that landscape. Initial information can be quickly put together from sources such as:

- product websites and marketing materials, and trade shows;

- projects funded by government research programs;

- policies outlining strategic directions and priorities in areas such as health care;

- academic articles, such as literature reviews;

- serious media outlets that reports on health, technology, etc.;

- information websites dedicated to the particular market sector;

- social networking and influencers; and

- market reports.

The earlier Challenge Areas chapters can be seen as environmental scans, highlighting priorities, projects, and products currently in the marketplace.

Once you have had a look around your Challenge Areas, you need to evaluate it and ask: What are the forces that are pushing or holding back innovation in the Challenge Areas, and are they the avenues for innovation?

PESTEL—Is an analysis of the macro-level forces that drive or constrain innovation or an organization: Political Economic Social Technological Environmental Legal (see https://blog. oxfordcollegeofmarketing.com/2016/06/30/pestel-analysis/). There is no single place to look for relevant information, but might include many of the sources mentioned previously, such as market reviews, policy briefs, and academic research.

SWOT—Was developed to look at individual organizations, but is also useful simple framework to evaluate a market sector (see https://www.mindtools.com/pages/article/newTMC_05. htm).

- **S**trengths—Where are existing solutions doing a good job, or where are emerging solutions likely to be strong?

- **W**eaknesses—Where are existing products not doing such a great job?

- **O**pportunities—What are the best or most likely directions for innovation?

- **T**hreats—What factors might be stopping innovation in this sector?

Looking at health services for people with dementia care, one could argue that the weaknesses outweigh the strengths. In Chapter 5, we noted that much of the care is provided in long-term residential settings, while community, services have often been under-resourced and under-developed. To some extent this could be seen as an opportunity to develop more preventative and community-based alternatives to long-term residential care that help people manage their own health conditions and live independently at home. However, a key threat is that effecting change and innovation in the health care sector has proved notoriously difficult.

15.8   BUSINESS MODEL CANVAS

Bringing all this together is the final step. The *Business Model Canvas* was developed by Alexander Osterwalder (Otsterwalder, Pigneur, and Clark, 2010) and is a simple template for developing new, or refining and documenting existing, business models. The Business Canvas is a visual chart describing a company's or product's value proposition, infrastructure, customers, and finances.

- A Business Canvas sets out the key ideas about how to reach your "market."

- Use the Business Canvas to distil your ideas in clear simple language.

- The Business Canvas is a key step in turning ideas into a "product."

The Business Canvas comprises nine building blocks, such as resources, key activities, customer segments and relationships, and market channels. It is probably a good idea to start with baby-steps and the *Lean Business Canvas* (Figure 15.1) can be seen as "Business Canvas Lite" and is more appropriate for early stage ideas about your product.

Lean Business Canvas		Team or Company Name: **Company/Project Name:**		Date: **[MM/DD/YY]**
The Problem	Your Solution	Value Proposition	Channels	Customer Groups
	Existing Solutions	Competitive Advantage	Metrics	

Figure 15.1: The lean business canvas template.

The boxes in the Lean Canvas will help you to create a basic business proposition and the various tools such as the Value Proposition, Environmental Scan, PESTEL, and SWOT, discussed above will help to fill out the following.

- **Problem**—What is your customer's key problem/aspiration?

- **Existing solutions**—What alternative products are in the marketplace already, or in the pipeline? What are their particular strengths and weaknesses? Your Environmental Scan will provide input here.

- **Your solution**—Describe the key feature(s) of your product and say how it will meet your customer's needs. How does it fit in the bigger picture?

- **Value proposition**—Why would the customer purchase or use your solution? What are the perceived benefits would they gain from your product?

- **Competitive advantage**—Say what sets your particular product apart for its competitors. What does it do better than the others?

- **Customer groups**—Who are your key target customers? Your personas and scenarios will be useful here. Who may be early adopters?

- **Channels**—How will you reach and inform your potential customers?

- **Metrics**—How will you measure and evaluate customer interest and usage of your product?

Box 15.2 Simple Music Player

Returning to the example of the Simple Music Player highlighted in Chapter 6, then we can create a very simple Lean Canvas as an illustration.

The Simple Music Player was taken to market by DesignAbility, a not-for-profit organization in the UK that aims to create and market attractive and usable assistive technologies (http://www.designability.org.uk/product/simple-music-player/). Working with DesignAbility was key to taking the prototype device to market, both in terms of final product development and the marketing of the device through the company E2L (http://www.e2l.uk.com). Sales of the Simple Music Player have now exceeded several thousand and it is available via distributors worldwide. Further information can be found at http://www.dementiamusic.co.uk. At this website there is information on how the device is making itself heard across the world, and it includes a video of a 96-year old Canadian man, clearly enjoying the music. The Simple Music Player is part of the product line at John Bell and Croyden, the flagship store for the Lloyds Pharmacy group in London.

Lean Business Canvas		Team or Company Name: **Simple Music Player**		Date: **[MM/DD/YY]**
The Problem Many people with dementia would like to listen to music, but they often need help from a caregiver to do this: • Devices are complicated to operate • Even where a person can operate a music-playing device, they often need encouragement. • Music is a personal thing and needs to be tailored to the person's preferences	**Your Solution** The Simple Music Player for people with dementia uses a single button to easily play music from uploaded MP3 music files. The device is an attractive retro style that can be immediately recognized as a music-playing device by a person with impaired cognitive abilities.	**Value Proposition** The Simple Music Player allows a person with moderate to severe dementia to access and listen to the music they enjoy. 	**Channels** The music player can be marketed to long-term care providers through their assistive technology suppliers. There are numerous channels to market directly to caregivers—online retailers, such as Amazon, pharmacies, and consumer electronics retailers who are increasingly interested in this sector. Advocacy groups and charities provide links to retailers for assistive devices and there are specialist online retailers such as alzstore.com.	**Customer Groups** Long-term residential care providers who wish to provide a more stimulating environment and enjoyable activities. Family caregivers who will purchase the music player to help the person with dementia to continue to listen to the music that they love and enjoy.
	Existing Solutions While there are many different consumer products in the marketplace (streaming, MP3 players, and even record players and radios), music-playing devices are typically too difficult or complicated to operate for a person with severe dementia.	**Competitive Advantage** While the idea may seem simple, there were no devices on the market that combined the style, simplicity, and robustness needed for this particular market	**Metrics** Interest in the product can be gauged by feedback and testimonials from potential and actual purchasers and end-users, for example at trade shows, and feedback from online sales. Growth in direct to consumer sales is the key metric.	

15.9 FIND OUT MORE

In Section IV, the "Reaching out" chapter in the book— Sixsmith, A. et al. (Eds.) (2020). *Knowledge, Innovation, and Impact: A Guide for the Engaged Health Researcher*, provides useful how-to chapters, case studies, and learning activities related to knowledge mobilization, commercialization, and communications.

Standard Business Canvas:

https://www.businessmodelsinc.com/about-bmi/tools/business-model-canvas/; and

A YouTube video on Business Model Canvas with examples: https://www.youtube.com/watch?v=-CakUeC1sCSs.

Lean Business Canvas:

https://www.lucidchart.com/blog/lean-canvas-model;

https://xtensio.com/how-to-create-a-lean-canvas/; and

You can watch this video on YouTube, which gives you some ideas about filling in the Lean Canvas. It uses Uber as an example: https://www.youtube.com/watch?v=pvIN9STpzCQ.

Stormboard is a free online collaboration tool that has pre-defined templates that group can work on simultaneously. They have empathy maps, the Lean and regular Business Canvas, business plans, persona development, etc. https://stormboard.com/.

CHAPTER 16

Emerging Issues and Future Directions

During the preparation of this book, the authors worked on various sections and contributed comments and shared feedback with each other. However, there were a number of emerging themes that we felt deserved some further discussion, especially in terms of future directions for research at the intersections of AgeTech, cognitive health, and dementia. We received some very useful feedback from reviewers of an early draft of the manuscript, which again highlighted key ideas and themes that warranted further exploration such as the notion of AgeTech as a concept in itself, the problematization of technology as a positive force for good, the need for stronger evidence on the outcomes of AgeTech in real-world environments. The manuscript for this book was being prepared at the very time that the coronavirus pandemic took hold in Europe and North America in the first half of 2020, and the longer-term impact of the pandemic on societies and economies worldwide is yet to unfold. Because of this, a conscious decision was made not try to integrate ideas relating to COVID-19 and AgeTech in the main body of the book and that a short section should focus on the issue in this final chapter. Future editions will be in a much better position to reflect on the impact of COVID-19 with more authority.

16.1 THE WORD "AGETECH"

In this book, we have used a neologism—**AgeTech**—to describe the use of existing and emerging technologies to meet the needs and aspirations of older adults, their families, and caregivers. While there is a long tradition of assistive devices for older adults, such as spectacles, walkers, prosthetic devices, home adaptations, etc., the focus of AgeTech is on utilizing existing and emerging advanced technologies, such as digital media, information and communication technologies (ICTs), mobile technologies, wearables, and smart home systems. AgeTech *products* include apps, devices, and systems, but also technology-enabled ser-

sdecoret: Shutterstock

vices, as well as knowledge products, such as information and government policy to support the use and adoption of technology. It is important to think of AgeTech as an **emerging sector within the wider digital economy**, rather than as a discrete set of clinical/social needs problems and engineering solutions that are specific to a particular group, recognizing that while technologies may be designed with older adults in mind, people of younger ages may also find them useful. This switch in perspective will help us to move away from a tech-driven approach to one that focuses on the commodification of emerging AgeTech solutions, the creation of "value" and exchange (market-based), distribution (rights and access), and the embedding of technology within social and caring relationships (social capital).

16.2   THE ROLE OF TECHNOLOGY

While technology is often seen as a solution to the problems and challenges of an increasingly aging population, especially in terms of health and social care needs and demands, technology should not be seen as, or promoted as, a substitute for human interaction. Instead, it is best framed as supporting human interaction, or as additional to existing face-to-face interaction, a facilitator rather than a replacement. Indeed, the economic argument that technology can be used to increase productivity, reduce costs, and increase capacity may be illusory. The real value proposition of new solutions should be to improve the range, reach, and quality of services and supports. Much of the narrative within AgeTech focuses on the problems, deficits, and impairments due to cognitive decline. As a counterbalance, a stronger emphasis is needed on the support of the aspirations, abilities, potential of individuals, and on their identity and sense of self As dementia care moves away from a focus on biomedical understandings to person-centered care, research in design and development for and with people with dementia has also shifted away from such a homogenous view of dementia that focuses on loss and disability. Furthermore, it could be mentioned here that the challenge for researchers has both significant individual

JOKE_PHATRAPONG: Shutterstock

and social ramifications, not least because of the many different forms of dementia and stages of the disease. The impact on the lived experience and the progression of the condition can vary greatly from person to person and their socio-cultural background. Further complexities in supporting dementia care is found in the wide variety of stakeholders involved and the varied context in which people with dementia live. These variables have broad implications for developing, designing, and evaluating technologies in a wide range of experiences that needs to be considered.

16.3   THE NEED FOR EVALUATION AND EVIDENCE ON THE OUTCOMES OF AGETECH

This book has described numerous projects and products designed to support cognitive health and dementia. However, one issue that emerged in the writing of this book was the need for more evidence to support wider adoption of AgeTech beyond the pilot or start-up stages such as research into the feasibility, the commercialization of and the evaluation of technologies in the real world. The projects and products described in earlier chapters certainly look promising and many have incorporated the design thinking discussed above. However, most do not go beyond talking about "possibilities" and "potential," and there are far fewer studies that describe the real-world impact. This challenge is increasingly being recognized and addressed. For example, the Centre for Aging and Brain Health Innovation (www.cabhi.com) has a funding program targeted at clinician–researcher partnerships to trial and evaluate innovative solutions at advanced stages of development. But it is important to look at how researchers can collaborate with industry or community to carry out the longer-term evaluation impact of interventions?

16.4   CAREGIVING, "GATEKEEPERS," AND CAREGIVERS

There is an increasing awareness that supporting people with dementia may primarily be achieved by technologies in support of the caregiving role. Providing care has become increasingly demanding due to the increased level and complexity of health conditions. Family caregivers often face challenges in juggling paid work, caregiving, and other family responsibilities. They need support and tools to provide care to ensure quality of life of their loved ones as well as themselves (Stall, 2019).

Caregiving role: Caregiving can be a source of stress, leading to depression and other medical problems. Caregivers often retreat from their wider social circles because of changes in the character and behavior of the people with dementia, resulting in social isolation (Schorch et al., 2016) However, we should not forget that caregiving may also be a source of satisfaction, pride, and sense of duty (Zarit, 2012). Given this complexity, Mayer and Zach (2013) argue that technologies aiming to enhance quality of life for people with dementia require a deep understanding of symptoms, problems, and user needs, not only concerning people with dementia, but also their families and caregivers.

Gatekeepers: The key decision-makers on purchasing and using the kinds of technologies described in this book are often family caregivers. This "gatekeeper" role is crucial in the design and commercialization of technologies as they need to see the value proposition of a suggested technology if it is to be incorporated into the lives of people with dementia. In many respects it is the caregiver should be seen as the customer and how technologies are perceived, appropriated, and integrated into care settings is a key research area.

Caregivers in the R&D process: While participatory design is feasible with people with dementia, it can be difficult to achieve. People with dementia, especially those in the moderate to severe stages of the disease, may not have the capacity to provide insightful feedback and may be confused or limited in their verbal communication. Successfully involving people with dementia in a research process can, therefore, necessitate considerable attention to building trust and engagement. In this context, caregivers can play a key role in the research and design process as a support or proxy for the person with dementia (Hellström et al., 2007; Karlsson et al., 2011). However, this is not without its challenges, adding pressure to the many demands on the caregiver. We should also recognize that their perspectives may not necessarily be congruent with the wishes or best interests of the people with dementia.

16.5   REFLECTIONS ON COVID-19

The COVID-19 pandemic has had a profound global impact, but recent evidence shows it has had disproportionate negative impacts on socially disadvantaged people and communities (Collaboration for Wellbeing and Health, 2020). It may also have initiated more difficulties for people with dementia and their caregivers as their everyday routines are disrupted, changes are needed in behavior (Van Bavel et al., 2020), and a lack of understanding of the disease or social distancing measures risks higher levels of exposure. Physical distancing and face-to-face contact were implemented in most countries to restrict infection, particularly among older adults where the risk of severe complications and subsequent mortality is much higher. The pandemic has also sharpened our focus on technology as a potential solution for supporting seniors and caregivers—*helping to connect people at a time of disconnection.* For example, the UK government pledged £750 million to ensure voluntary, community and social enterprises can continue their essential work during the COVID-19 pandemic, including £200 million for the Coronavirus Community Support Fund (Gov.uk, 2020). The Canadian federal government announced in April 2020 a $350 million fund for community organizations that support vulnerable groups, including older adults, and specifically referred to using technology to replace in-person, face-to-face contact and social gatherings with virtual contact through telephone, texts, teleconferences, or the Internet. There appears to be increasing awareness of the importance of technology among older adults themselves. A poll commissioned by AGE-WELL NCE found that some three-quarters of older Canadians thought that technology can help them to live independently, stay active and healthy, and reduce social isolation. Anecdotally, we are seeing many more older adults using technology, belying the stereotype that they are intrinsically technophobic.

It is possible to highlight some key areas that have emerged under the COVID-19 pandemic where AgeTech may play a role:

Long-term residential care: The inadequacies of care home systems worldwide have been laid bare by the pandemic: high levels of infection, serious illness, and death; painful separation from families who are no longer able to visit; debilitating stress and fear among workers; and inadequate resources, poor levels of training, and ineffective management. Recent decades have seen long-term residential homes increasingly cater to people with dementia and, while technology will never be a panacea for these systemic problems, AgeTech has to be part of a long overdue serious conversation and shake-up of the sector.

Social isolation and mental health: COVID-19 has highlighted the issue of isolated seniors living in our communities. As we saw in earlier chapters, even before the COVID-19 crisis, about 19% of older adults experience social isolation (National Seniors Council, 2014), which in turn can lead to stress, depression, and poor health outcomes. The issue of physical distancing, isolation and mental health has now been identified as a priority for health research. For example, preliminary results of a survey by the Canadian Institute of Aging identified a number of key areas for research: impact of prolonged social distancing on mental health; short- and long-term impacts on physical and mental health of those recovered; impact of social isolation on older adults, family members and those that are dying; impact of inactivity and social isolation; role of technology in supporting communication with older adults; and virtual care delivery models.

Access-for-all: Our societies and economies have started, more than ever, to rely on technology-based ways of working, caring, and connecting socially. Lack of access to devices and the Internet and barriers to using technology—the digital divide between the technology haves and have-nots—has now become a life and death issue. There have been numerous ad hoc initiatives by tech retailers, network providers, and local government and community organizations, to help people to become digitally connected. However, governments worldwide need to recognize that citizenship and social and economic participation in the 21st century has to mean digital participation and more needs to be done to ensure this as a basic right for everyone. This is not just a matter of infrastructure or ensuring connection—devices, apps and websites need to be easier to use and we need better community-based supports to help people engage with technology. For example, government funding to support community organizations to mobilize local IT expertise and volunteers would be a practical, ground-up way of enabling people to help each other.

Using what we already have: The ideas of *telehealth* and *telecare* are not new, but have been painfully slow to be adopted on a large scale by health agencies. Under the pandemic lockdown, we have seen these suddenly become more readily available as a practical and convenient way of providing basic health advice and care. Health care service providers and funders need to be much more proactive in adopting the kinds of technology-based products that have been described in this book. The role of AgeTech more broadly should be extended as a means of supporting aging in place and improving care even after the pandemic passes.

fizkes: Shutterstock

16.6   CONCLUSION

We have explored the ways in which research and development of technologies can improve the cognitive health and quality of life of older adults and people with dementia, their families and carers. We have identified several areas in which AgeTech can benefit both individuals and societies with the proviso that such developments are designed "with" rather than "for" older adults and with participatory approaches including AgeTech in the home, for services and in the community, technology that supports autonomy and independence, connectedness, mobility, healthy lifestyles, and finances. This book also gives guidance on ethical working to support participation in AgeTech, policy development, and ensuring real-world impact.

It is inevitable that a book such as this one will emphasize the positive benefits and outcomes of AgeTech. But we also need to be aware of the realities as well—that living with cognitive decline can be a very challenging experience for many older adults and their caregivers. Nevertheless, AgeTech has huge potential to improve health and quality of life as we age, to enable older adults to enjoy life and to create businesses and jobs and generate economic growth despite the longer-term implications of COVID-19 that are still to emerge. We should learn from the challenges of the pandemic and let them be the catalyst for meaningful change for the better in the lives of older adults.

AgeTech Glossary

Unless otherwise stated, descriptions of the glossary terms are sourced from the text.

A

Actor Model (Stakeholder expertise map): Is usually a table that contains a complete list of stakeholder types, the specific stakeholder, their role, and their particular goals/priorities. The model should be something that is jointly created and refined by the stakeholders involved so that there is buy-in from everyone concerned.

AgeTech: AgeTech refers to the use of advanced technologies such as information and communication technologies (ICTs), robotics, mobile technologies, artificial intelligence (AI), ambient systems, and pervasive computing to drive technology-based innovation to benefit older adults (Pruchno, 2019).

AGE-WELL NCE: Is Canada's federally funded AgeTech research and innovation network. Launched in 2015, the network has grown to 250 researchers at 42 universities and research centers across Canada, with over 100 technologies, services, policies, and practices being developed across the network. See https://agewell-nce.ca/.

Age-related cognitive changes: Are relatively minor changes in cognitive abilities as we get older.

Assistive devices: Enable people with dementia to continue to live in their own homes and communities. These include assistive technologies that help older adults by monitoring the environment and aiding with simple tasks. A variety of products aiming to accomplish this already exist on the market, such as, wayfinding technologies, GPS locators that monitor the location of individuals outdoors and motion sensors as well as cameras to use within their homes (Bossen et al., 2015).

B

Brain health: Encompasses both the structural and physical health of the brain in addition to the cognitive functioning component.

Business Model Canvas: Was developed by Alexander Osterwalder (Otsterwalder, Pigneur, and Clark, 2010) and is a simple template for developing new, or refining and documenting existing, business models. The Business Canvas is a visual chart describing a company's or product's value proposition, infrastructure, customers, and finances. The chart sets out key ideas about how to reach your "market," and is a key step in turning ideas into a "product."

C

Challenge Areas: A challenge is an important, but difficult and complex problem area that demands innovation and the application of real-world solutions. A challenge is engaging and is worthwhile to pursue because it: results in significant social and economic benefits to older adults; is difficult to accomplish, but is ultimately solvable; requires collaboration across many disciplines and groups; pushes the scientific envelope; must capture popular imagination and political support; inspires hope; and brings people together to work for the common good.

Co-creation: Is about involving customers, end-users, and other stakeholders in stages in the development of a new product or service. These stakeholders are not merely providers of information—they are also active contributors and collaborators within a research team. Most importantly, they are the experts on their own experiences and situations.

Co-creation workshops: Are workshops where researchers and stakeholders can work together to articulate problems, identify priorities, and map out high-level solutions.

Co-creators: Co-creators are stakeholders that work together to articulate problems, identify priorities, and map out high-level solutions.

Cognitive health: Concerns the ability to perform cognitive mental processes, such as learning, intuition, attention, judgment, language, and memory. It can be greatly affected by age-related, debilitating disorders, such as dementia, that significantly reduce a person's quality of life and ability to live independently. It is related to having a healthy brain, which in turn is related to cardiovascular, physical, and hormonal health, but there is likely to be a large variation among individuals in terms of cognitive function irrespective of brain health. As well the concept of cognitive health encompasses our ability to successfully perform the wide variety of mental processes our brains are able to accomplish.

Commercialization: Is the process of bringing a new product or service to market and someone will have to "productize," manufacture, package, market, and sell the device, all of which needs financial investment (Battersby and Viswanathan, 2020).

Community-based participatory research approach (CBPR): Its purpose is to develop opportunities for meaningful exchange of knowledge using less conventional ways of working (e.g., mapping workshops, informal dialogue sessions, knowledge cafés) to enable participation of people with varying degrees of knowledge, expertise, and skills in research. As an approach it was applied using a variety of methods (i.e., in-depth interviews, photo-voice, community mapping, deliberative dialogue, feedback forums, and knowledge cafés) to help not only identify key stakeholders but also co-create solutions to enable the successful transition of older adults into affordable housing.

Co-researching: This approach integrates participants in research such as older adults and people with dementia as part of the research team as trained researchers who, for example, will conduct interviews with participants also diagnosed with dementia or be involved in the analytical process. They can also be involved in project design and dissemination. Co-researching is grounded on a participatory and emancipatory philosophy as part of action research, but has reportedly demonstrated more personal benefits for participants with dementia (Tanner, 2012).

Crystallized intelligence: Represent the skills and abilities we have learned over the lifespan, such as vocabulary and general knowledge, remains stable through the later decades of life (Harada, Love, and Triebel, 2014).

Customer groups: Are the key target customers in the commercialization process.

D

Deliberative dialogue: Builds on dialogue and consensus-building techniques, enabling participants to work together (often with expert input) to develop an agreed view or set of recommendations. As participants may then be involved in taking their recommendations forward to decision-makers, this can encourage shared responsibility for implementation. Examples include national dialogues on science and technology. Source: https://www.involve.org.uk/resources/knowledge-base/what/deliberative-public-engagement.

Dementia: Is used as an umbrella term that describes a group of cognitive symptoms that can be induced by pathological changes to the structure and function of the brain, such as Alzheimer's disease, stroke, or traumatic brain injury. Dementia affects many cognitive processes, most notably are declines in memory, attention, reasoning, and communication abilities. It is regarded as one of the main causes of disability later in life, ahead of cancer, cardiovascular disease, and stroke (Chambers, Bancej, and McDowell, 2016).

Diagnostic and Statistical Manual of Mental Disorders: Is an internationally accepted classification of mental disorders, with the aim of improving diagnoses, treatment, and research. Now in its fifth edition and is published by the American Psychiatric Association.

Digital divide: Where certain groups of individuals, such as people with dementia, are disempowered and sometimes excluded from the benefits of accessing or using a technology (Fang et al., 2019). It is found within most societies, especially for many people who are marginalized.

Digital walking maps: Can help people find their ways to and from places, as well as using public transportation. Such maps should include pictures to help people recognize familiar places, while voice interfaces can help to reassure people and prompt them to find their way.

Digital storytelling: Is an expansion … of telling stories by using technology—in various media forms such as text, photos, sounds and video—to reimage a story, pass down personal history, and share experiences social connectedness and general well-being. It can create opportunities for older adults to reflect on their life histories and share their narrative, building new connections and relationships (Waycott et al., 2013). Additionally, digital storytelling helps to interconnect generations, a personal digital story could be passed down between generations and retold many times (Hausknecht, Vanchu-Orosco, and Kaufman, 2019).

E

E-Health (or Telehealth): Allows people to connect with health services to monitor their health conditions or receive services at home.

Electronic health records (EHR): An electronic health record (EHR) is a secure, integrated collection of a person's encounters with the health care system; it provides a comprehensive digital view of a patient's health history. Source: Canada Health Infoway: https://www.infoway-inforoute.ca/en/solutions/digital-health-foundation/electronic-health-record.

Environmental scan: Surveys and interprets relevant data to identify external opportunities and threats that could influence future decisions. …and…should be used as part of the strategic planning process. Components of external scanning that could be considered include: trends; competition; technology; customers; economy; labor supply; political/legislative arena. Source: Society for Human Resource Management: https://www.shrm.org/resourcesandtools/tools-and-samples/hr-qa/pages/basics-of-environmental-scanning.aspx.

Executive functioning: Is a set of mental skills that include working memory, flexible thinking, and self-control. We use these skills every day to learn, work, and manage daily life. Trouble with executive function can make it hard to focus, follow directions, and handle emotions, among other things. Source: Understood.org: https://www.understood.org/en/learning-thinking-differences/child-learning-disabilities/executive-functioning-issues/what-is-executive-function.

F

Financial wellness: Is about the overall financial health of a person, sometimes referred to as financial security. This is an important component of healthy aging (Sixsmith et al., 2014). Financial wellness, as a component of healthy aging, affects many areas of a person's life.

Fluid intelligence: One's ability to problem solve or process new information tends to slowly decline after age 30. Fluid cognition encompasses a wide range of cognitive, such as processing speed, attention, memory, verbal fluency, visuospatial abilities, and other executive functions.

G

Gatekeepers: These can include: health service providers; family members and carers; management (of institutions); and ethics board/committee members. Gatekeepers often have the decision-making power over the people they are gatekeeping and thus can decide whether they participate and/or how they engage in the research. They are a "stakeholder group." As well they can be found in: community-based organizations (e.g., dementia care services), faith groups, memory clinics, and caregiver support groups.

H

Health-related technologies: A health technology is the application of organized knowledge and skills in the form of devices, medicines, vaccines, procedures, and systems developed to solve a health problem and improve quality of lives. Source: World Health Organization: https://www.who.int/health-technology-assessment/about/healthtechnology/en/.

Healthy lifestyle: At a population level, healthy living refers to the practices of population groups that are consistent with supporting, improving, maintaining, and/or enhancing health. As it applies to individuals, healthy living is the practice of health-enhancing behaviors, or put simply, living in healthy ways. Source: Government of Canada: https://www.canada.ca/en/public-health/services/health-promotion/healthy-living.html.

I

Independence: Refers to functional ability (regarding both activities of daily living and instrumental activities of daily living) in everyday life, remaining active and contributing to the community (Dwyer and Gray, 2000), with financial security and social and psychological resilience. This relates to the notion of maintaining autonomy, defined as "the right of the individual to be self-determining and to make independent decisions about his or her life" (Kanny and Slater, 2008, p. 194).

Information and communication technologies (ICTs): Are mobile technologies, wireless networks and the ubiquitous Internet.

Internet of Medical Things: The development and interconnection of: medical grade and health centered devices; individuals; caregivers; and professional health care providers (Gatouillat et al., 2018).

The Internet of Things (IoT): Is the concept that multiple devices or smart devices can be interconnected to communicate via the Internet both with other devices and the user (Dey et al., 2018). Some examples of these devices include: voice-activated assistants; fitness trackers; smart thermostats; media systems; cleaning devices; and even refrigerators.

K

Knowledge mobilization (KM): Is about connecting the right people with the right products and information (Simeonov, Kobayashi, and Grenier, 2017). Terms like KM, knowledge translation, and knowledge exchange are often used interchangeably (Barwick et al., 2014), but they all focus on the flow of knowledge, ideas, and innovation between researchers and stakeholders with the aim of driving change and ultimately social and/or economic impact. Increasingly, the common view of knowledge translation as a one-way linear flow of information from knowledge producers to consumers is seen as too simplistic. KM should be seen as a dynamic and integrated activity as discussed in earlier chapters on co-creation.

M

Major Neurocognitive Disorder: Is to replace the commonly used term dementia in its revised classification (American Psychiatric Association, 2014). It highlighted different etiologies (or types), such as Major Neurocognitive Disorder due to Alzheimer's disease.

Memory: The ability to recall names and places is the cognitive function that most declines with age.

Metrics: Refers to how will you measure and evaluate customer interest and usage of your product.

Mild cognitive impairment (MCI): Refers to the relatively minor but noticeable cognitive changes that may be a precursor to dementia. It impacts cognitive health in a more mild form and it is defined as a decline in cognition greater than what is expected for normal aging, but does not interfere with activities of daily living (Petersen et al., 1999)."

Mild Neurocognitive Disorder: Mild neurocognitive disorder goes beyond normal issues of aging. It describes a level of cognitive decline that requires compensatory strategies and accommodations to help maintain independence and perform activities of daily living. To be diagnosed with this disorder, there must be changes that impact cognitive functioning. Source: www.psychiatry.org: file:///C:/Users/Owner/AppData/Local/Temp/APA_DSM-5-Mild-Neurocognitive-Disorder.pdf.

Mock-up: A rudimentary model of a device.

N

Normal Cognitive Decline: Cognitive abilities tend to decline a little as we age, and we refer to this as normal cognitive decline. We typically experience changes in our cognitive abilities and our ability to remember things, work out problems, and carry out tasks declines a little.

P

Paper prototype: Simple cut-out paper models of new devices to show what they might look like to potential users.

Persona and scenario: Short, semi-fictional case studies of target customer/user and their living situations and problems. Personas and scenarios are particularly useful in illustrating and highlighting the human side of what it is to live with dementia.

Products as discussed by AGE-WELL NCE: Technology products—Systems and devices aimed at directly supporting the people with dementia and caregivers; service products—These are the delivery models and mechanisms that will allow new technologies and solutions to actually be provided to the user or patient; Policy and practice—They are knowledge products, such as information, policy briefs, guidelines, standards and regulations, models of good practice, as well as health-related information for the public.

PESTEL (Political Economic Social Technological Environmental Legal): Is an analysis of the macro-level forces that drive or constrain innovation or an organization.

S

Serviceable Obtainable Market (SOM): What a company could realistically reach in the short medium and longer terms.

Social companion robots: Mimic the presence of a pet, like a dog, cat, or baby seal or a person. They are designed to assist in decreasing the feelings of social isolation and loneliness many older adults experience, but can be modified to become a social companion as well as a telepresence for caregiver information (Vaughn, Shaw, and Molloy, 2015).

Social participation: Is about being able to remain actively engaged within society, the economy, and within our families and communities (Woolrych et al., 2019). Being involved in volunteering, work, leisure, religious and cultural activities, as well as civic engagement may improve the well-being of older adults and increase the social and human capacity of their communities.

Stakeholders: Are those who also have vested interest, in the area being addressed, e.g., people experiencing cognitive decline, informal caregivers, service providers, and government.

Storyboard: A simple visual and written/verbal narrative of how a device, system or service might work in practice.

Sustainable Development Goals (SDGs): Address the key challenges facing all countries, such as poverty, inequality, and climate change. The aim is to achieve them by 2030 in order to "leave no one behind" (United Nations Sustainable Goals, n.d.). Aging is seen as an important aspect of the SDG agenda: Preparing for an aging population is vital to the achievement of the integrated 2030 Agenda, with aging cutting across the goals on poverty eradication, good health, gender equality, economic growth and decent work, reduced inequalities and sustainable cities. The approach is about rights, and how all older adults, even those with dementia, can and should be enabled to be participants in society.

SWOT: Was developed to look at individual organizations, but is also a useful and simple framework to evaluate a market sector. The following are components of SWOT: Strengths—where are existing solutions doing a good job, or where emerging solutions are likely to be strong?; Weaknesses—where are existing products not doing such a great job; Opportunities—what are the best or most likely directions for innovation?; Threats—what factors might be stopping innovation in this sector?

T

Technology Readiness Levels (TRL): Is an approach by NASA to technology (Mankins, 2009), and it encompasses nine levels of maturity with level one concerning ideas and level nine being technology in its most mature form. A simplified model is the Product Innovation Pathway (Sixsmith et al., 2020) that describes various broad steps: initial ideas, project planning, development, implementation and evaluation, and commercialization. The technologies involved include smart home systems, assistive robots, socially connected platforms, serious games, and smart connected age-friendly cities and communities.

Telecare: Includes remote monitoring, home security, device controls, and emergency response systems that enhance the safety and security, independence of older adults, and their ability to stay living at home and in their community.

Theater method: This involves actors performing a scripted enactment of how a new product (as well as situation, relationship or scenario) might operate—this is particularly useful for demonstrating the technology or service to an audience of stakeholders and getting feedback.

Total Addressable Market (TAM): In terms of dementia, this refers to the total number of people with dementia can be referred to for a dementia-related product.

U

User: The user is the person or organization who will use or operate the product or the system.

V

Value proposition: A value proposition is part of the development of the Business Canvas Model which is a brief outline of your ideas about your product. It needs to clearly say what the need or problem is that you are addressing and who the intended customer is. The crucial aspect of this is the "value," the key benefits that the customer will get from buying or using the product: Can be used as the basis for outreach and publicity.

W

Wicked problem: Refers to a challenge as an important, but difficult and complex problem area that has resisted solution through traditional single-discipline research approaches.

Wizard-of-Oz technique: A rudimentary working mock-up of a device that is manipulated behind the scenes by an operator to interact with the person using it.

The World Café methodology: Generates in-depth discussions, by engaging a large group of cross-sectoral stakeholders in concurrent smaller group dialogues, in order to explore a single question or use a progressively deeper line of inquiry through several conversational rounds (Brown, Isaacs, and World Café Company, 2002).

References

AARP Public Policy Institute. (2011). How the travel patterns of older adults are changing: Highlights from the 2009 national household travel survey: https://assets.aarp.org/rgcenter/ppi/liv-com/fs218-transportation.pdf. 44

AGE-WELL NCE (2019). The future of technology and aging research in Canada [booklet]: https://agewell-nce.ca/wp-content/uploads/2019/01/Booklet_8_Challenges_English_5_final_PROOF_rev.pdf. 4

AGE-WELL NCE (2020). Workpackages—Core research projects listing: https://agewell-nce.ca/workpackages-core-research-projects-listing. 24, 68

AGE-WELL News (2020a). Alert system aims to keep people with dementia safe. *AGE-WELL News*: https://agewell-nce.ca/archives/7589. 25

AGE-WELL News (2020b). Braze Mobility unveils obstacle-detection system for wheelchairs: https://agewell-nce.ca/archives/5855. 47

AGE-WELL News (2020c). A new technology opens up a new world. AGE-WELL News: https://agewell-nce.ca/archives/8221. 61

Albert, M. S., Jones, K., Savage, C. R., Berkman, L., Seeman, T., Blazer, D., and Rowe, J. W. (1995). Predictors of cognitive change in older persons: MacArthur studies of successful aging. *Psychology and Aging*, 10(4): 578–589: https://www.alz.org/media/Documents/alzheimers-facts-and-figures.pdf. 8

Alzheimer's Association and Centers for Disease Control and Prevention (2013). The health brain initiative: the public health road map for state and national partnerships, 2013–2018. Chicago, IL: Alzheimer's Association: https://www.cdc.gov/aging/pdf/2013-healthy-brain-initiative.pdf [Report].

Alzheimer's Association (2018). What is dementia? [Fact sheet]: https://www.alz.org/alzheimers-dementia/what-is-dementia. 10, 13

Alzheimer's Association (2020). Alzheimer's disease facts and figures: special report on the front lines: primary care physicians and Alzheimer's care in America. Washington, D.C.: Alzheimer's Association: https://www.alz.org/media/Documents/alzheimers-facts-and-figures.pdf. 9

142 REFERENCES

Alzheimer's Disease International (n.d.). Dementia plans: https://www.alz.co.uk/ dementia-plans. 101

Alzheimer's Disease International publication team (2018). From plan to impact: Progress toward targets of the global action plan on dementia. London (UK): Alzheimer's Disease International: https://www.alz.co.uk/adi/pdf/from-plan-to-impact-2018.pdf. 100

Alzheimer Europe (2013). 2013: The prevalence of dementia in Europe: https://www.alzheimer-europe.org/Policy/Country-comparisons/2013-The-prevalence-of-dementia-in-Europe. 101

Alzheimer Europe (2018). 2018: Comparison of national dementia strategies in Europe: https://www.alzheimer-europe.org/Policy-in-Practice2/Country-comparisons/2018-Comparison-of-National-Dementia-Strategies.

Alzheimer's Society of Canada (2016). Prevalence and monetary costs of dementia in Canada: http://alzheimer.ca/sites/default/files/files/national/statistics/prevalenceandcostsofdementia_en.pdf.

Alzheimer Society of Canada (2017). The Canadian Alzheimer's Disease and Dementia Partnership: http://alzheimer.ca/sites/default/files/files/national/advocacy/caddp_strategic_objectives_e.pdf. 104

American Psychiatric Association (2014). Coding update. Diagnostic and Statistical Manual of Mental Disorders, 5th ed. [Supplement]: https://dsm.psychiatryonline.org/pb-assets/dsm/update/DSM5CodingUpdateICD9Archive.pdf. 13, 136

American Psychiatric Association (2020). *Diagnostic and Statistical Manual of Mental Disorders (DSM–5)*: https://www.psychiatry.org/psychiatrists/practice/dsm. 13

Arntzen, C., Holthe, T., and Jentoft, R. (2016). Tracing the successful incorporation of assistive technology into everyday life for younger people with dementia and family carers. *Dementia* 15(4): 646–662. DOI: 10.1177/1471301214532263. 17

Asghar, I., Cang, S., and Yu, H. (2019). An empirical study on assistive technology supported travel and tourism for the people with dementia. *Disability and Rehabilitation: Assistive Technology*, 1–12. DOI: 10.1080/17483107.2019.1629119. 47

Astell, A. J., Czarnuch, S., and Dove, E. (2018). System development guidelines from a review of motion-based technology for people with dementia or MCI. *Frontiers in Psychiatry*, 9: 189. DOI: 10.3389/fpsyt.2018.00189. 81

Astell, A. and Fels, D. (2015–2020). Understanding the needs of older adults (WP1 NEEDS-OA): tools for user needs gathering to Support Technology Engagement (1.2 TUNGSTEN). [Project overview]: https://forum.agewell-nce.ca/index.php/1.2_TUNGSTEN:Main. 81

Barrett, L. (2014). Home and Community Preferences of the 45+ Population 2014. [Report] https://www.aarp.org/content/dam/aarp/research/surveys_statistics/il/2015/home-community-preferences. DOI: 10.26419/res.00105.001.

Bartlett, R. (2012). Modifying the diary interview method to research the lives of people with dementia. *Qualitative Health Research*, 22: 1717–1726. DOI: 10.1177/1049732312462240. 79

Bartlett, H. and Martin, W. (2002). Ethical issues in dementia care research. In Wilkinson, H. (Ed.), *The Perspectives of People with Dementia: Research Methods and Motivations*. London, UK: Jessica Kingsley Publishers, pp. 47–62. 11

Barwick, M., Phipps, D., Myers, G., Johnny, M., and Coriandoli, R. (2014). Knowledge translation and strategic communications: unpacking differences and similarities for scholarly and research communications. *Scholarly and Research Communication*, 5(3): 0305175: http://src-online.ca/index.php/src/article/view/175/345. DOI: 10.22230/src.2014v5n3a175. 115, 136

Battersby, L., Fang, M. L., Canham, S. L., Sixsmith, J., Moreno, S., and Sixsmith, A. (2017). Co-creation methods: informing technology solutions for older adults. In Jia, Z. and Salvendy, G. (Eds.), *Human Aspects of IT for the Aged Population. Aging, Design and User Experience*, pp. 77–89. New York: Springer. DOI: 10.1007/978-3-319-58530-7_6. 71

Battersby, L. and Viswanathan, P. (2020). Commercializing research innovations: an introduction for researchers. In Sixsmith, A., Sixsmith, J., Fang, M. L., and Mihailidis, A. (Eds.) (2020). *Knowledge, Innovation and Impact in Health: A Guide for the Engaged Researcher*. New York: Springer. 115, 132

Benbow, S. M. and Kingston, P. (2014). "Talking about my experiences at times disturbing yet positive": producing narratives with people living with dementia. *Dementia*. Advance online publication. DOI: 10.1177/1471301214551845. 79

Benke, E., producer (2020). People fixing the world: gyms for life: https://www.bbc.com/news/av/stories-51389184/the-town-paying-for-people-over-65-to-get-fit [film]. 53

Blackman, S., Matlo, C., Bobrovitskiy, C., Waldoch, A., Fang, M. L., Jackson, P., Mihailidis, A., Nygårdm, L., Astell, A., and Sixsmith, A. (2016). Ambient assisted living technologies for aging well: a scoping review. *Journal of Intelligent Systems*, 25(1): 55–69. DOI: 10.1515/jisys-2014-0136. 24

Blackman, T., Mitchell, L., Burton, E., Jenks, M., Parsons, M., Raman, S., and Williams, K. (2003). The accessibility of public spaces for people with dementia: a new priority for the 'open city', *Disability and Society*, 18(3): 357–371. DOI: 10.1080/0968759032000052914. 11

Boman, I.-L., Nygård, L., and Rosenberg, L. (2014). Users' and professionals' contributions in the process of designing an easy-to-use videophone for people with dementia. *Disability and Rehabilitation: Assistive Technology*, 9(2): 164–172. DOI: 10.3109/17483107.2013.769124. 79

Bould, E., McFadyen, S., and Thomas, C. (2019). Dementia-friendly sport and physical activity guide. London: Alzheimer's Society: https://www.alzheimers.org.uk/sites/default/files/2019-02/19003-Sports_Leisure_guide_online.pdf. 54

Bossen, A., Kim, H., Steinhoff, A., Strieker, M., and Williams, K. (2015). Emerging roles for telemedicine and smart technologies in dementia care. *Smart Homecare Technology and TeleHealth*, 2015(3): 49–57. DOI: 10.2147/SHTT.S59500. 16, 131

Brooks, J., Savitch, N., and Gridley, K. (2017). Removing the 'gag': involving people with dementia in research as advisers and participants. *Social Research Practice*, 3: 3–14. 71, 72, 78

Brown, J., Isaacs, D., and World Café Company (2002). *The World Café: Shaping Our Futures through Conversations that Matter*. Mill Valley, CA: Whole Systems Associates. 76 , 139

Burns, A. and Zaudig, M. (2002). Mild cognitive impairment in older people. *Lancet*, 360(9349): 1963–1965. DOI: 10.1016/S0140-6736(02)11920-9. 12

Canada Mental Health Association (n.d.). The most common Cognitive Disorders include: Delirum and Dementia (such as Alzheimer's). Canada Mental Health Association Fort Frances Branch: http://www.cmhaff.ca/cognitive-disorders.

Canadian Mental Health Association (n.d.). *Cognitive Disorders*. 10

Canadian Alzheimer's Society (2017). *The Canadian Alzheimer's Disease and Dementia Partnership*. [Booklet]: http://alzheimer.ca/sites/default/files/files/national/advocacy/caddp_strategic_objectives_e.pdf. 101

Capstick, A. and Ludwin, K. (2015). Place memory and dementia: Findings from participatory film-making in long-term social care. *Health and Place*, 34: 157–163. DOI: 10.1016/j.healthplace.2015.05.012. 79

Centers for Disease Control and Prevention (n.d.). Cognitive impairment: A call for action, now! [Brochure]: https://www.cdc.gov/aging/pdf/cognitive_impairment/cogimp_poilicy_final.pdf. 99

Centers for Disease Control and Prevention (2009). What is a healthy brain? New research explored perceptions of cognitive health among diverse older adults. [Fact sheet]: https://www.cdc.gov/aging/pdf/perceptions_of_cog_hlth_factsheet.pdf. 13, 51

Centers for Disease Control and Prevention (2011). The CDC healthy brain initiative: progress 2006–2011: https://www.cdc.gov/aging/pdf/HBIBook_508.pdf. 99

Chambers, L. W., Bancej, C., and McDowell, I. (Eds.) (2016). Prevalence and monetary costs of dementia in Canada. Toronto: The Alzheimer's Society of Canada; Ottawa: The Public Health Agency of Canada: http://alzheimer.ca/sites/default/files/files/national/statistics/prevalenceandcostsofdementia_en.pdf. 99, 133

Chignell, M. and Liu, L. (2020). Workpackage 6: Technology for maintaining good mental and cognitive health: ICT applications for screening, assessment and interventions to enhance mental health—Workpackage 6.1 MEN-ASSESS: https://agewell-nce.ca/research/research-themes-and-projects/wp6-new. 54

Cleveland Clinic (2019). 6 Pillars of brain health. [Website]: https://healthybrains.org/pillars/. 13

Cobigo, V., Potvin, L. A., Fulford, C., Chalghoumi, H., Hanna, M., Plourde, N., and Taylor W. D. (2019). A conversation with research ethics boards about inclusive research with persons with intellectual and developmental disabilities. In Cascio, M. A. and Racine, E. (Eds.), *Research Involving Participants with Cognitive Disability and Differences: Ethics, Autonomy, Inclusion, and Innovation,* 185–196. Oxford: Oxford University Press. DOI: /10.1093/oso/9780198824343.003.0016. 89

Collaboration for Wellbeing and Health (2020). What can we do to help those already facing disadvantage, in the COVID-19 outbreak? The Health Foundation. [Webpage]: https://www.health.org.uk/newsletter-feature/what-can-we-do-to-help-those-already-facing-disadvantage-in-the-covid-19. 128

Cooney, M., Pihl, J., Larsson, H., Orand, A., and Aksoy, E. E. (2019). Exercising with an "Iron Man": Design for a robot exercise coach for persons with dementia: https://arxiv.org/abs/1909.12262. 53

Cooper, C., Procter, P., and Penders, J. (2016). Dementia and robotics: people with advancing dementia and their carers driving an exploration into an engineering solution to maintaining safe exercise regimes. In Sermeus, W., Procter, P., and Webster, P. (Eds.), Nursing informatics 2016, eHealth for all: entry level collaboration—from project to realization. *Studies in Health Technology and Informatics* (225). Amsterdam: IOS Press, 545–552: http://shura.shu.ac.uk/13601/2/ __staffhome.hallam.shu.ac.uk_STAFFHOME2_2_hwbcc2_MyWork_Work-%20Personal_SHTI225-0545.pdf. 53

Creavin, S. T., Wisniewski, S., Noel-Storr, A. H., Trevelyan, C. M., Hampton, T., Rayment, D., Thom, V. M., Nash, K. J. E., Elhamoui, H., Milligan, R., Patel, A. S., Tsivos, D. V., Wing, T., Phillips, E., Kellman, S. M., Shackleton, H. L., Singleton, G. F., Neale, B. E.,

Watton, M. E., and Cullum, S. (2016). Mini-mental State Examination (MMSE) for the detection of dementia in clinically unevaluated people aged 65 and over in community and primary care populations. *Cochrane Database of Systematic Reviews*, (1): https://www.cochranelibrary.com/cdsr/doi/10.1002/14651858.CD011145.pub2/full. DOI: 10.1002/14651858.CD011145.pub2. 89

Daum, C. H., Neubauer, N., Oliva, C., Beleno, R., and Liu, L. (2019). Accuracy and usability of a mobile alert system for community citizens to locate persons with dementia who get lost. *Innovation in Aging*, 3(Suppl 1), S453– S454. DOI: 10.1093/geroni/igz038.1698. 25

dementia Australia™. (n.d. a). Dementia-friendly home app: https://www.dementia.org.au/information/resources/technology/dementia-friendly-home-app. 24

dementia Australia™. (n.d. b). EDIE—Educational Dementia Immersive Experience: https://www.dementia.org.au/information/resources/technology/edie. 32

Dempsey, L., Dowling, M., Larkin, P., and Murphy, K. (2016). Sensitive interviewing in qualitative research. *Research in Nursing and Health*, 39(6): 480–490. DOI: 10.1002/nur.21743. 90, 92

designability (2020). Product case study: Simple Music Player: https://designability.org.uk/ projects/products/simple-music-player/. 40

Dey, N., Hassanien, A. E., Bhatt, C., Ashour, A. S., and Satapathy, C. S. (Eds.) (2018). Internet of things and big data analytics toward next-generation intelligence. *Studies in Big Data Book*, 30. Cham, CH: Springer. DOI: 10.1007/978-3-319-60435-0. 18, 136

Douglass-Bonner, A. and Potts, H. W. W. (2013). Exergame efficacy in clinical and non-clinical populations: a systematic review and meta-analysis. Presented at the *6th World Congress on Social Media, Mobile Apps, Internet/Web 2.0, London, UK*. (2013). 17

Drageset, I. (2019). Research involving older persons with dementia; ethical considerations, *Klinisk Sygepleje*. 33(03): 219–233. DOI: 10.18261/issn.1903-2285-2019-03-05. 87, 91

Droplet® (n.d.). https://www.droplet-hydration.com/introducing-droplet-hydration/.

Dwyer, M. and Gray, A. (2000). Maintaining independence in old age: policy challenges. *Social Policy Journal of New Zealand*, 13: 83–94. 35, 135

E2L Consultancy Group (2015). Welcome to The E2L Consultancy Group. [Website]: http://www.e2l.uk.com. 40

Family Caregiver Alliance. (2018). Caregiver statistics: demographics: https://www.caregiver.org/caregiver-statistics-demographics. 101

Fang, M. L., Woolrych, R., Sixsmith, J., Canham, S., Battersby, L., and Sixsmith, A. (2016). Place-making with older persons: establishing sense-of-place through participatory community mapping workshops. *Social Science and Medicine*, 168: 223–229. DOI: 10.1016/j.socscimed.2016.07.007. 12, 21

Fang, M. L., Woolrych, R., Sixsmith, J., Canham, S. L., Battersby, L., Ren, T. H., and Sixsmith, A. (2017). Place-making with seniors: toward meaningful affordable housing, final report. Vancouver: STAR Institute, Simon Fraser University. 78

Fang, M. L., Siden, E., Korol, A., Demesthias, M. A., Sixsmith, J., and Sixsmith, A. (2018). Exploring the intended and unintended consequences of eHealth on older people: a health equity impact assessment. *Human Technology*, 14(3): 297–323. DOI: 10.17011/ht/urn.201811224835. 1

Fang, M. L., Canham, S. L., Battersby, L., Sixsmith, J., Wada, M., and Sixsmith, A. (2019). Exploring privilege in the digital divide: implications for theory, policy, and practice. *The Gerontologist*, 59(1): e1–e15. DOI: 10.1093/geront/gny037. 95, 133

Federal/Provincial/Territorial Ministers Responsible for Seniors Forum (2012). Age-friendly workplaces: promoting older worker participation. *Gatineau: Federal/Provincial/Territorial Ministers Responsible for Seniors Forum* [Report]: https://www.canada.ca/content/dam/esdc-edsc/documents/corporate/seniors/forum/promoting.pdf. 68

Feinberg, L., Reinhard, S., Houser, A., and Choula, R. (2011). Valuing the invaluable: 2011—Update the growing contributions and costs of family caregiving. Washington, D.C.: AARP Public Policy Institute: https://assets.aarp.org/rgcenter/ppi/ltc/i51-caregiving.pdf. 44

Foley, S., Pantidi, N., and McCarthy, J. (2019). Care and design: an ethnography of mutual recognition in the context of advanced dementia. In *Proceedings of the 2019 CHI Conference on Human Factors in Computing Systems – CHI '19*. Presented at the 2019 CHI Conference. Glasgow: ACM Press, pp. 1–15. DOI: 10.1145/3290605.3300840.

Freed, S. A., Ross, L. A., and Stavrinos, D. (2019). Older adults' opinions on different vehicle technologies. *Innovation in Aging*, 3(Supp 1): S342–S343. DOI: 10.1093/geroni/igz038.1241. 47

Fujisoft Inc. (2018). PALRO GARDEN. [Website]. 17

Garvelink, M. M., Jones, C. A., Archambault, P. M., Roy, N., Blair, L., and Légaré, F. (2017). Deciding how to stay independent at home in later years: development and acceptability testing of an informative web-based module. *JMIR Human Factors*, 4(4): e32. DOI: 10.2196/humanfactors.8387. 35

Gatouillat, A., Badr, Y., Massot, B., and Sejdic, E. (2018). Internet of medical things: a review of recent contributions dealing with cyber-physical systems in medicine. *IEEE Internet of Things Journal*, 5(5): 3811–3822. DOI: 10.1109/JIOT.2018.2849014. 18, 135

Geda, Y. E. (2012). Mild cognitive impairment in older adults. Research support, N.I.H., Extramural research support, Non-U.S. Goverment Review. *Current Psychiatry Reports*, 14(4): 320–327. DOI: 10.1007/s11920-012-0291-x. 12

Göbel, C. and Zwick, T. (2010). Which personnel measures are effective in increasing productivity of old workers?, *ZEW Discussion Papers*, no. 10-069. Mannheim: Zentrum für Europäische Wirtschaftsforschung (ZEW): https://www.econstor.eu/bitstream/10419/41432/1/636380688.pdf. 69

Government of Canada (2017). Dementia in Canada, including Alzheimer's disease: highlights from the Canadian chronic disease surveillance system: https://www.canada.ca/en/public-health/services/publications/diseases-conditions/dementia-highlights-canadian-chronic-disease-surveillance.html. 99

Government of Canada (2018). Promoting the labour force participation of older Canadians. Ottawa: Employment and Social Development Canada: https://www.canada.ca/en/employment-social-development/corporate/seniors/forum/labour-force-participation.html. 66

Gov.uk (2020). Financial support for voluntary, community and social enterprise (VCSE) organisations to respond to coronavirus (COVID-19): https://www.gov.uk/guidance/financial-support-for-voluntary-community-and-social-enterprise-vcse-organisations-to-respond-to-coronavirus-covid-19. 128

Grönvall, E. and Kyng, M. (2013). On participatory design of home-based healthcare. *Cognition, Technology, and Work*, 15(4), 389–401: https://link.springer.com/article/10.1007/s10111-012-0226-7?shared-article-renderer. DOI: 10.1007/s10111-012-0226-7. 84

Grustam, A. S. (2014). Cost-effectiveness of telehealth interventions for chronic heart failure patients: a literature review. *International Journal of Technology Assessment in Health Care*, 30(1): 59–68. DOI: 10.1017/S0266462313000779. 109

Gschwind, Y. J., Eichberg, S., Ejupi, A., de Rosario, H., Kroll, M., Marston, H. R., Drobics, M., Annegarn, J., Wieching, R., Lord, S. R., Aal, K., Vaziri, D., Woodbury, A., Fink, D., and Delbaere, K. (2015). ICT-based system to predict and prevent falls (iStoppFalls): results from an international multicenter randomized controlled trial. *European Review of Aging and Physical Activity*. Act. 12. DOI: 10.1186/s11556-015-0155-6.

Hadjistavropoulos, T., Taati, B., Prkachin, K., and Marchildon, G. (2018). WP6.3—PAIN-AS-SESS: development, implementation and evaluation of an automated pain detection system for older adults with dementia: https://agewell-nce.ca/age-well-core-research-projects. 30

Hampson, C. and Morris, K. (2018). Research into the experience of dementia: methodological and ethical challenges. *Journal of Social Sciences and Humanities*, 1(1): 15–19: http://insight.cumbria.ac.uk/id/eprint/3805/. 87, 89

Harada, C. N., Natelson Love, M. C., and Triebel, K. (2014). Normal cognitive aging. *Clinical Geriatric Medicine*, 29(4): 737–752: https://www.ncbi.nlm.nih.gov/pmc/articles/PMC4015335/. DOI: 10.1016/j.cger.2013.07.002. 8, 133

Hausknecht, S., Vanchu-Orosco, M., and Kaufman, D. (2019). Digitising the wisdom of our elders: connectedness through digital storytelling. *Ageing and Society*, 39(12): 2714–2734. DOI: 10.1017/S0144686X18000739. 61, 134

Hearthstone. (2017). Home4Care. Home4Care. http://home4care.org/. 32

Hedman, R., Hellström, I., Ternestedt, B.-M., Hansebo, G., and Norberg, A. (2016). Sense of self in Alzheimer's research participants. *Clinical Nursing Research*, 27(2): 1–22. DOI: 10.1177/1054773816672671. 87

Hellström, I., Nolan, M., Nordenfelt, L., and Lundh, U. (2007). Ethical and methodological issues in interviewing persons with dementia. *Nursing Ethics*,14(5): 608–619. DOI: 10.1177/0969733007080206. 75, 128

Henderson, C., Knapp, M., Fernández, J. L., Beecham, J., Hirani, S. P., Cartwright, M., Rixon, L., Beynon, M., Rogers, A., Bower, P., Doll, H., Fitzpatrick, R., Steventon, A., Bardsley, M., Hendy, J., Newman, S. P., and Whole System Demonstrator evaluation team. (2013, March 20). Cost effectiveness of telehealth for patients with long term conditions (Whole Systems Demonstrator telehealth questionnaire study): nested economic evaluation in a pragmatic, cluster randomised controlled trial. 346: j2065. *BMJ*: https://www.ncbi.nlm.nih.gov/pubmed/23520339. DOI: 10.1136/bmj.f1035. 110

Hendriks, N., Huybrechts, L., Wilkinson, A., and Slegers, K. (2014, October). Challenges in doing participatory design with people with dementia. In *Proceedings of the 13th Participatory Design Conference: Short Papers, Industry Cases, Workshop Descriptions, Doctoral Consortium Papers, and Keynote Abstracts-Volume 2* (pp. 33-36). 78, 83, 84

Hodge, J., Balaam, M., Hastings, S., and Morrissey, K. (2018). Exploring the design of tailored virtual reality experiences for people with dementia. *Proceedings of the 2018 CHI Conference on Human Factors in Computing Systems – CHI '18*. ACM Press, 1–13. DOI: 10.1145/3173574.3174088. 17

Hodge, J., Montague, K., Hastings, S., and Morrissey, K. (2019). Exploring media capture of meaningful experiences to support families living with dementia. *Proceedings of the 2019 CHI Conference on Human Factors in Computing Systems—CHI '19*. Presented at the 2019 CHI Conference. Glasgow: ACM Press, pp. 1–14. DOI: 10.1145/3290605.3300653. 60

Hogeweyk Dementia Village (2017). https://hogeweyk.dementiavillage.com/en/. 25

Holthe, T., Jentoft, R., Arntzen, C., and Thorsen, K. (2018). Benefits and burdens: family caregivers' experiences of assistive technology (AT) in everyday life with persons with young-onset dementia (YOD). *Disability and Rehabilitation: Assistive Technology*, 13: 754–762. DOI: 10.1080/17483107.2017.1373151. 17

Hope4Care (2017). Hope4Care: http://home4care.org/.

INDUCT (Interdisciplinary Network for Dementia Using Current Technology) (n.d.). ESRs Projects: https://www.dementiainduct.eu/projects/. 55

InterRAI. (2020). https://www.interrai.org.

Ip, E. H., Barnard, R., Marshall, S. A., Lu, L., Sink, K., Wilson, V., Chamberlain, D., and Rapp, S. R. (2017). Development of a video-simulation instrument for assessing cognition in older adults. *BMC Medical Informatics and Decision Making*, 17, Article no. 161: https://bmc-medinformdecismak.biomedcentral.com/articles/10.1186/s12911-017-0557-7. DOI: 10.1186/s12911-017-0557-7. 30

Jagosh, J., Bush, P. L., Salsberg, J., Macaulay, A. C., Greenhalgh, T., Wong, G., Cargo, M., Green, L. W., Herbert, C. P., and Pluye, P. (2015). A realist evaluation of community-based participatory research: partnership synergy, trust building and related ripple effects. *BMC Public Health*, 15(1), Article no. 725. DOI: 10.1186/s12889-015-1949-1. 72, 74

Jawaid, S. and McCrindle, R. (2015). Computerised help information and interaction project for people with memory loss and mild dementia. *Journal of Pain Management*, 9(3): 265–276. 38

Jenkins, N., Keyes, S., and Strange, L. (2016). Creating vignettes of early onset dementia: An exercise in public sociology. *Sociology*, 50: 77–92. DOI: 10.1177/0038038514560262. 79

Jog, M., Patel, R., Duval, C., Frank, J., and Teasell, R. (2015–2020). An in-home intelligent exercise system for physical rehabilitation, enhancing musculoskeletal function, and preventing adverse events (5.3 IIES-PHYS): https://agewell-nce.ca/age-well-core-research-projects. 54

Kanny, E. M. and Slater, D. Y. (2008). Ethical reasoning. In: Schell, B., and Schell, J., Eds., *Clinical and Professional Reasoning in Occupational Therapy*. Philadelphia: Wolters Kluwer/Lippincott and Wilkins, pp. 188–226. 35, 135

Karlsson, E., Axelsson, K., Zingmark, K., and Sävenstedt, S. (2011). The challenge of coming to terms with the use of a new digital assistive device: a case study of two persons with mild dementia. *The Open Nursing Journal*, 5: 102–110. DOI: 10.2174/1874434601105010102. 128

Karlsson, E., Zingmark, K., Axelsson, K., and Sävenstedt, S. (2017). Aspects of self and identity in narrations about recent events: communication with Individuals individuals with Alzheimer's disease enabled by a digital photograph diary. *Journal of Gerontological Nursing*, 43(6): 25–31. DOI: 10.3928/00989134-20170126-02. 18

Khosravi, P., Rezvani, A., and Wiewiora, A. (2016). The impact of technology on older adults' social isolation. *Computers in Human Behavior*, 63: 594–603. DOI: 10.1016/j.chb.2016.05.092. 58

Kim, K. and Antonopoulos, R. (2011). Unpaid and paid care: the effects of child care and elder care on the standard of living. Levy Economics Institute of Bard College. Working paper no. 691: http://www.levyinstitute.org/pubs/wp_691.pdf. DOI: 10.2139/ssrn.1946377. 65

Kim, S. K. and Park, M. (2017). Effectiveness of person-centered care on people with dementia: a systematic review and meta-anlaysis. *Clinical Intervention in Aging*, 12: 381–397. DOI: 10.2147/CIA.S117637. 27

Kitwood, T. (1997). *Dementia Reconsidered: The Person Comes First*. Berkshire, UK: Open University Press. DOI: 10.1136/bmj.318.7187.880a. 11

Knoefel, F., Wallace, B., Goubran, R., Sabra, I., and Marshall, S. (2019). Semi-autonomous vehicles as a cognitive assistive device for older adults. *Geriatrics*, 4(4): 63. DOI: 10.3390/geriatrics4040063. 47

Kong, A. P.-H. (2015). Conducting cognitive exercises for early dementia with the use of apps on iPads. *Communication Disorders Quarterly*, 36(2): 102–106. DOI: 10.1177/1525740114544026. 18

Krick, T., Huter, K., Domhoff, D., Schmidt, A., Rothgang, H., and Wolf-Ostermann, K. (2019). Digital technology and nursing care: a scoping review on acceptance, effectiveness and efficiency studies of informal and formal care technologies. *BMC Health Services Research*, 19, Article no. 400. DOI: 10.1186/s12913-019-4238-3. 109

Kroemeke, A. and Gruszczynska, E. (2016). Well-being and institutional care in older adults: cross-sectional and time effects of provided and received support. *PloS ONE*, 11(8): e0161328. DOI: 10.1371/journal.pone.0161328. 27

Kubiak, M. (2016). Silver economy-opportunities and challenges in the face of population ageing. *European Journal of Transformation Studies*, 4(2): 18–38. 1

Lavalliere, M., Burstein, A. A., Arezes, P., and Coughlin, J. F. (2016). Tackling the challenges of an aging workforce with the use of wearable technologies and the quantified-self. *Dyna rev. fac.nac.minas* [Online]. 83(197): 38–43. DOI: 10.15446/dyna.v83n197.57588. 68

Lee, J. and Hirdes, J. (2020). Data-driven decision-making in health care. AGE-WELL NCE Inc. [Webpage]: https://forum.agewell-nce.ca/index.php/7.3_3DHC:Main. 30

Leroi, I., Woolham, J., Gathercole, R., Howard, R., Dunk, B., Fox, C., O'Brien, J., Bateman, A., Poland, F., Bentham, P., Burns, A., Davies, A., Forsyth, K., Gray, R., Knapp, M., Newman, S., McShane, R., and Ritchie, C. (2013). Does telecare prolong community living in dementia? A study protocol for a pragmatic, randomized controlled trial. *Trials*, 14, Article no. 349. DOI: 10.1186/1745-6215-14-349. 110

Lindsay, S., Brittain, K., Jackson, D., Ladha, C., Ladha, K., and Olivier, P. (2012, May). Empathy, participatory design and people with dementia. In *Proceedings of the SIGCHI Conference on Human Factors in Computing Systems*. Paris: ACM Press, pp. 521–530. DOI: 10.1145/2207676.2207749. 78, 84

Liu, L., Stroulia, E., Nikolaidis, I., Miguel-Cruz, A., and Rincon, A. R. (2016). Smart homes and home health monitoring technologies for older adults: A systematic review. *International Journal of Medical Informatics*, 91: 44–59. DOI: 10.1016/j.ijmedinf.2016.04.007. 109

Longeway, S. (2019, October 7). Ethical by design: new framework aims to avoid unintentional consequences of technology, *Waterloo Stories*. University of Waterloo: https://uwaterloo.ca/stories/global-impact/ethical-design. 96

Lothian, K. and Philp, I. (2001). Maintaining the dignity and autonomy of older people in the healthcare setting. *BMJ* (Clinical research ed.), 322(7287): 668–670. DOI: 10.1136/bmj.322.7287.668.

Lubberink, R., Blok, V., Van Ophem, J., and Omta, O. (2017). Lessons for responsible innovation in the business context: A systematic literature review of responsible, social and sustainable innovation practices. *Sustainability*, 9(5): 721. DOI: 10.3390/su9050721. 1

Lunn, K., Sixsmith, A., Lindsay, A., and Vaarama, M. (2003). Traceability in requirements through process modelling, applied to social care applications. *Information and Software Technology*, 45(15): 1045–1052. DOI: 10.1016/S0950-5849(03)00132-0. 74

Maisen, N., Pitts, K., Pudney, K., and Zachos, K. (2018). Evaluating the use of daily care notes software for older people with dementia. *International Journal of Human–Computer Interaction*, 35(7): 605–619. DOI: 10.1080/10447318.2018.1482052. 15

Mankins, J. (2009). Technology readiness levels: a retrospective. *Acta Astronautica*, 65(9–10): 1216–1223. DOI: 10.1016/j.actaastro.2009.03.058. 116

Martyn-Hemphill, A., producer (2020). BBC "Old Age Cities" [film] BBC World Hacks: https://www.bbc.com/news/av/stories-47490097/how-japan-is-helping-pensioners-stay-happy-and-have-fun.

Martin, S., Kelly, G., Kernohan, W. G., McCreight, B., and Nugent, C. (2008). Smart home technologies for health and social care support. *Cochrane Database of Systematic Reviews*, Issue 4, Article no. CD006412. DOI: 10.1002/14651858.CD006412.pub2. 109

MacDonald, A. and Cooper, B. (2007). Long-term care and dementia services: an impending crisis. *Age and Ageing*, 36(1): 16–22. DOI: 10.1093/ageing/afl126. 11

Mayer, J. M. and Zach, J. (2013). Lessons learned from participatory design with and for people with dementia. *MobileHCI 2013 Proceedings of the 15th International Conference on Human-Computer Interaction with Mobile Devices and Services*, August 27–30; Munich, Germany. pp.: 540–545. DOI: 10.1145/2493190.2494436. 127

McKeown, J., Clarke, A., Ingleton, C., and Repper, J. (2010). Actively involving people with dementia in qualitative research. *Journal of Clinical Nursing*, 19(13–14): 1935–1943. DOI: 10.1111/j.1365-2702.2009.03136.x. 72, 75

Mehta, S. (2004). The emerging role of academia in commercializing innovation. *Nature Biotechnology*, 22(1): 21–24. DOI: 10.1038/nbt0104-21. 115

Meiland, F. J., Bouman, A. I., Sävenstedt, S., Bentvelzen, S., Davies, R. J., Mulvenna, M. D., … Bengtsson, J. E. (2012). Usability of a new electronic assistive device for community-dwelling persons with mild dementia. *Aging & Mental Health*, 16(5), 584–591. DOI: 10.1080/13607863.2011.651433. 84

Meiland, F., Innes, A., Mountain, G., Robinson, L., Roest, H., García-Casal, J. A., Gove, D., Thyrian, J. R., Evans, S., Dröes, R-M., Kelly, F., Kurz, A., Casey, D., Szcześniak, D., Dening, T., Craven, M. P., Span, M., Felzmann, H., Tsolaki, M., and Franco-Martin, M. (2017). Technologies to support community-dwelling persons with dementia: a position paper on issues regarding development, usability, (cost-) effectiveness, deployment and ethics. *JMIR Rehabilitation and Assistive Technologies*, 4(1): e1. DOI: 10.2196/rehab.6376. 15

Michaud, F. and Negat, G. (2020). Mobile robotics for activities of daily living assistance—Workpackage 3.1 VIGIL: https://agewell-nce.ca/research/research-themes-and-projects/wp3-new. 37, 38

Miller, B. and Mitchell, I. (2020). Collaborative power mobility for an aging population—Workpackage 3.2 CoPILOT: https://agewell-nce.ca/research/research-themes-and-projects/wp3-new. 46

Millett, S. (2011). Self and embodiment: A bio-phenomenological approach to dementia. *Dementia*, 10(4): 509–522. DOI: 10.1177/1471301211409374. 87

Morin, K. B. (2018, March 2). Effects of technological innovations on social interactions. Samuel Centre for Social Connectedness: https://www.socialconnectedness.org/effects-of-technological-innovations-on-social-interactions/.

Morrison, C., Cutrell, E., Dhareshwar, A., Doherty, K., Thieme, A., and Taylor, A. (2017, October). Imagining artificial intelligence applications with people with visual disabilities using tactile ideation. In *Proceedings of the 19th International ACM SIGACCESS Conference on Computers and Accessibility* (pp. 81-90). 17

Moyle, W., Jones, C., Murfield, J., Thalib, L., Beattie, E., Shum, D., and Draper, B. (2017a). Using a therapeutic companion robot for dementia symptoms in long-term care: reflections from a cluster-RCT. *Aging and Mental Health*, 23(3): 329–336. DOI: 10.1080/13607863.2017.1421617. 17

Moyle, W., Jones, C. J., Murfield, J. E., Thalib, L., Beattie, E., Shum, D., O'Dwyer, S. T., Mervin, M. C., and Draper, B. M. (2017b). Use of a robotic seal as a therapeutic tool to improve dementia symptoms: a cluster-randomized controlled trial. *Journal of the American Medical Directors Association*, 18(9): 766–773. DOI: 10.1016/j.jamda.2017.03.018. 15

Murphy, J., Gray, C. M., van Achterberg, T., Wyke, S., and Cox, S. (2010). The effectiveness of the Talking Mats framework in helping people with dementia to express their views on well-being. *Dementia*, 9(4): 454–472. DOI: 10.1177/1471301210381776. 62, 79

Murphy, J., Holmes, J., and Brooks, C. (2017). Measurements of daily energy intake and total energy expenditure in people with dementia in care homes: the use of wearable technology. *The Journal of Nutrition, Health and Aging*, 21(8): 927–932. DOI: 10.1007/s12603-017-0870-y. 53

NAC and AARP Public Policy Institution. (2015). Caregiving in the U.S. Research Report: https://www.aarp.org/content/dam/aarp/ppi/2015/caregiving-in-the-united-states-2015-report-revised.pdf. 44

Nagarajan, R. N., Wada, M., Fang, M. L., and Sixsmith, A. (2019). Defining the contribution of relevant mechanisms to sustaining an ageing workforce: A bibliometric review. *European Journal of Ageing*, 16(3): 337–361. DOI: 10.1007/s10433-019-00499-w. 66

National Academies of Sciences, Engineering, and Medicine; Institute of Medicine, Board on Health Sciences Policy, Forum on Neuroscience and Nervous System Disorders, Bain, L., and Norris, S. P. (2016). Assessing the impact of application of digital health records on Alzheimer's disease research. Workshop summary (2016). Washington, D.C.: The National Academies Press. DOI: 10.17226/21827. 18

National Institute on Aging (2017). Reducing risks to cognitive health: https://www.nia.nih.gov/health/cognitive-health-and-older-adults. 13

National Seniors Council (2014). Report on the Social Isolation of Seniors: https://www.canada.ca/en/national-seniors-council/programs/publications-reports/2014/social-isolation-seniors/page03.html. 129

Nijhof, N., van Gemert-Pijnen, L. J., Woolrych, R., and Sixsmith, A. (2013). An evaluation of preventative sensor technology for dementia care. *Journal of Telemedicine and Telecare*, 19(2): 95–100. DOI: 10.1258/jtt.2012.120605. 110

Novalte (2020). emitto. [Website]: https://www.novalte.ca/emitto/. 25

OA-INVOLVE (2019). Older adults' engagement: engaging people living with dementia in technology project, barriers and facilitators, solutions and tips: http://www.oa-involve-agewell.ca/uploads/1/2/7/2/12729928/dementia_brief_oct_17_2019.pdf. 94

Ogonowski, C., Aal, K., Vaziri, D., Rekowski, T. V., Randall, D., Schreiber, D., Wieching, R., and Wulf, V. (2016). ICT-based fall prevention system for older adults: qualitative results from a long-term field study. *ACM Transactions on Computer-Human Interaction*, 23: 1–33. DOI: 10.1145/2967102. 53

Orpwood, R., Sixsmith, A, Torrington, J., Chadd, J., Gibson, G., and Chalfont, G. (2007), Designing technology to support quality of life of people with dementia. *Technology and Disability*, 19(2): 103–112. DOI: 10.3233/TAD-2007-192-307. 40

Osterwalder, A., Pigneur, Y., and Clark, T. (2010). *Business Model Generation: A Handbook for Visionaries, Game Changers, and Challengers*. Strategyzer series. Hoboken, NJ: John Wiley and Sons. 121, 131

Oxley, J., Charlton, J., Logan, D., O'Hern, S., Koppel, S., and Meuleners, L. (2019). Safer vehicles and technology for older adults. *Traffic Injury Prevention*, 20(Supp 2): S176–S179. DOI: 10.1080/15389588.2019.1661712. 47

Petersen, R. C., Smith, G. E., Waring, S. C., Ivnik, R. J., Tangalos, E. G., and Kokmen, E. (1999). Mild cognitive impairment clinical characterization and outcome. Archives of Neurology, 56(3): 303–308. DOI: 10.1001/archneur.56.3.303. 12, 136

Petersen, R. C. (2004). Mild cognitive impairment as a diagnostic entity. *Journal of Internal Medicine*, 256(3): 183–194. DOI: 10.1111/j.1365-2796.2004.01388.x.

Petersen, R. C. (2011). Clinical practice. Mild cognitive impairment. *The New England Journal of Medicine*, 364(23), 2227. DOI:10.1056/NEJMcp0910237. 12

Phillipson, L. and Hammond, A. (2018). More than talking: A scoping review of innovative approaches to qualitative research involving people with dementia. *International Journal of Qualitative Methods*, 17(1): https://journals.sagepub.com/doi/10.1177/1609406918782784. DOI: 10.1177/1609406918782784. 79

Pierce, R. (2010). Complex calculations: ethical issues in involving at-risk healthy individuals in dementia research. *Journal of Medical Ethics*, 36(9): 553–557. DOI: 10.1136/jme.2010.036335. 94

Plamondon, K. M., Bottorff, J. L., and Cole, D. C. (2015). Analysing data generated through deliberative dialogue: Bringing knowledge translation into qualitative analysis. *Qualitative Health Research*, 25(11):1529–1539. DOI: 10.1177/1049732315581603. 76

Pruchno, R. (2019). Technology and aging: An evolving partnership. The Gerontologist, 59(1): 1–5. DOI: 10.1093/geront/gny153. 1, 131

Public Health Agency of Canada (2017). Dementia in Canada, including Alzheimer's disease: highlights from the Canadian chronic disease surveillance system: https://www.canada.ca/en/public-health/services/publications/diseases-conditions/dementia-highlights-canadian-chronic-disease-surveillance.html.

Public Health Agency of Canada (2018). National dementia conference: inspiring and informing a national dementia strategy for Canada: https://www.canada.ca/en/services/health/publications/diseases-conditions/national-dementia-conference-report.html. 101

Rendover Inc. (2019). [Website]. https://rendever.com/. 54

Robinson, L., Brittain, K., Lindsay, S., Jackson, D., and Olivier, P. (2009). Keeping In Touch Everyday (KITE) project: developing assistive technologies with people with dementia and their carers to promote independence. *International Psychogeriatrics*, 21(3): 494–502. DOI: 10.1017/S1041610209008448. 80

Rodrigues, E., Christie, G., Cosco, T., Farzan, F., Sixsmith, A., and Moreno, S. (In press). Healthy Cognitive Aging—and how it can be assessed. *Perspectives on Psychological Science*. 12

Rogers, W. A. and Mitzner, T. L. (2017). Envisioning the future for older adults: autonomy, health, well-being, and social connectedness with technology support. *Futures*, 87: 133–139. DOI: 10.1016/j.futures.2016.07.002. 37

Rule, L. (2010). Digital storytelling: never has storytelling been so easy or so powerful. *Knowledge Quest*, 38(4): 56–57: https://www.lib.uwo.ca/cgi-bin/ezpauthn.cgi?url=http://search.proquest.com/docview/609381695?accountid=15115. 61

Sachs-Ericsson, N. and Blazer, D. G. (2015). The new DSM-5 diagnosis of mild neurocognitive disorder and its relation to research in mild cognitive impairment. *Aging Mental Health*. 19(1): 2–12. DOI: 10.1080/13607863.2014.920303. 13

Salazar, R., Velez, C. E., and Royall, D. R. (2014). Telephone screening for mild cognitive impairment in hispanics using the Alzheimer's questionnaire. *Experimental Aging Research*, 40(2): 129–139. DOI: 10.1080/0361073X.2014.882189. 17

Salthouse, T. A. (2010). Selective review of cognitive aging. *Journal of the International Neuropsychological Society*, 16(5): 754–760. DOI: 10.1017/S1355617710000706. 8

Savoy, T., (2018, February 26). Technology that can foster aging in place. *Washington Post*. [Newspaper]: https://tinyurl.com/y2krg8u9. 24

Saxon, L., Ebert, R., and Sobhani, M. (August 13, 2019). Health impacts of unlimited access to networked transportation in older adults. *The Journal of mHealth*. https://thejournalofmhealth.com/health-impacts-of-unlimited-access-to-networked-transportation-in-older-adults/. 48

Schorch, M., Wan, L., Randall, D. W., and Wulf, V. (2016). Designing for those who are overlooked—Insider perspectives on are practices and cooperative work of elderly informal caregivers. In *CSCW '16: Proceedings of the 19th ACM Conference on Computer-Supported Cooperative Work and Social Computing*, Gergie, D., and Morris, M. R. (Chairs). New York: ACM Press, 785–797. DOI: 10.1145/2818048.2819999. 127

Schrum, M., Park, C. H., and Howard, A. (2019). Humanoid therapy robot for encouraging exercise in dementia patients. In *2019 14th ACM/IEEE International Conference on Human-Robot Interaction (HRI)*. 11–14 March 2019. Daegu, Korea (South), Korea (South). DOI: 10.1109/HRI.2019.8673155. 53

Segkouli, S., Paliokas, I., Tzovaras, D., Lazarou, I., Karagiannidis, C., Vlachos, F., and Tsolaki, M., (2018). A computerized test for the assessment of mild cognitive impairment subtypes in sentence processing. *Aging, Neuropsychology, and Cognition*, 25(6): 829–851. DOI: 10.1080/13825585.2017.1377679. 15, 17

Simeonov, D., Kobayashi, K., and Grenier, A. (2017). A resource for knowledge mobilization in aging and technology. [Internet] Toronto: AGE-WELL NCE: http://agewell-nce.ca/research/crosscutting-activities. 115, 136

Sixsmith, A. (2006). New technologies to support independent living and quality of life for people with dementia. *Alzheimer's Care Quarterly*, 7(3): 194–202. 15, 95, 102

Sixsmith, A. and Gibson, G. (2007). Music and well-being in dementia. *Ageing and Society*, 27(1): 127–146. DOI: 10.1057/9780230287624_5. 40

Sixsmith, A., Gibson, G., Orpwood, R., and Torrington, J. (2007a). Developing a technology "wish list" to enhance the quality of life of people with dementia. *Gerontechnology*, 6(1): 2–19. DOI: 10.4017/gt.2007.06.01.002.00. 39

Sixsmith, A., Hine, N., Nield, I., Clarke, N., Brown, S., and Garner, P. (2007b) Monitoring the well-being of older people. *Topics in Geriatric Rehabilitation*, 23(1): 9–23. DOI: 10.1097/00013614-200701000-00004. 23

Sixsmith, A. and Sixsmith, J. (2008). Ageing in place in the United Kingdom. *Ageing International*, 32(3): 219–235. DOI: 10.1007/s12126-008-9019-y. 21, 106

Sixsmith, A., Orpwood, R., and Torrington, J. (2010). Developing a music player for people with dementia. *Gerontechnology*, 9(3): 421–427. DOI: 10.4017/gt.2010.09.03.004.00. 15, 40

Sixsmith, A. (2013). Technology and the challenge of aging. In: Sixsmith, A. and Gutman, G. (Eds.), *Technologies for Active Aging*. New York: Springer, pp. 7–26. DOI: 10.1007/978-1-4419-8348-0_2. 3

Sixsmith, A., Mihailidis, A., and Simeonov, D. (2017). Aging and technology: taking the research into the real world. *Public Policy and Aging Report*, 27(2): 74–78. DOI: 10.1093/ppar/prx007. 1, 21

Sixsmith, A., Sixsmith, J., Fang, M. L., and Mihailidis, A., (2020). *Knowledge, Innovation, and Impact in Health—A Guide for the Engaged Researcher*. Basel: Springer. https://www.springer.com/gp/book/9783030343897. 116, 138

Sixsmith, A. (In press). AgeTech: technology-based solutions for aging societies. In Rootman, I., Edwards, P., Levasseur, M., and Grunberg, F. (Eds.), *Promoting the Health of Older Adults: The Canadian Experience* (n.p.). Toronto: Canadian Scholars' Press.

Sixsmith, J., Sixsmith, A., Fänge, A. M., Naumann, D., Kucsera, C., Tomsone, S., and Woolrych, R. (2014). Healthy ageing and home: the perspectives of very old people in five European countries. *Social Science and Medicine*, 106: 1–9. DOI: 10.1016/j.socscimed.2014.01.006. 65, 134

Sixsmith, J., Fang, M. L., Woolrych, R., Canham, S. L., Battersby, L., and Sixsmith, A. (2017). Ageing well in the right place: partnership working with older people. *Working with Older People*, 21(1): 40–48. DOI: 10.1108/WWOP-01-2017-0001. 1, 106

Smeddinck, J. D., Herrlich, M., and Malaka, R., (2015). Exergames for physiotherapy and rehabilitation: a medium-term situated study of motivational aspects and impact on functional reach. In *Proceedings of the 33rd Annual ACM Conference on Human Factors in Computing Systems – CHI '15*. Presented at the 33rd Annual ACM Conference, ACM Press, Seoul, Republic of Korea, pp. 4143–4146. DOI: 10.1145/2702123.2702598. 17

Smith, K. J. and Victor, C. (2019). Typologies of loneliness, living alone and social isolation, and their associations with physical and mental health. *Ageing and Society*, 39(8): 1709–1730. DOI: 10.1017/S0144686X18000132. 57

Snelling, S. (2019, September 18). Study finds ridesharing services improve older adults' lives. *Next Avenue*: https://www.nextavenue.org/ridesharing-services-improve-lives/. 48

Span, M., Hettinga, M., Vernooij-Dassen, M., Eefsting, J., and Smits, C. (2013). Involving people with dementia in the development of supportive IT applications: a systematic review. *Ageing Research Reviews*, 12(2): 535–551. DOI: 10.1016/j.arr.2013.01.002. 78

Spencer, E. (2017). The use of deception in interpersonal communication with Alzheimer's disease patients. *The Midwest Quarterly*, 58(2): 176–194: https://tinyurl.com/y5wjta5c. 16

Stall, N. (2019). We should care more about caregivers. *Canadian Medical Association Journal*, 191(9): E245-E246. DOI: 10.1503/cmaj.190204. 127

Suijkerbuijk, S., Nap, H. H., Cornelisse, L., IJsselsteijn, W. A., de Kort, Y. A., Minkman, M. M., and Baglio, F. (2019). Active involvement of people with dementia: a systematic review of studies developing supportive technologies. *Journal of Alzheimer's Disease*, 69(4): 1041–1065. DOI: 10.3233/JAD-190050. 80

Suter, E., Arndt, J., Arthur, N., Parboosingh, J., Taylor, E., and Deutschlander, S. (2009). Role understanding and effective communication as core competencies for collaborative practice. *Journal of Interprofessional Care*, 23(1): 41–51. DOI: 10.1080/13561820802338579. 75

Taha, J., Czaja, S. J., and Sharit, J. (2016). Technology training for older job-seeking adults: The efficacy of a program offered through a university-community collaboration. *Educational Gerontology*, 42(4): 276–287: https://www.tandfonline.com/doi/full/10.1080/03601277.2015.1109405. DOI: 10.1080/03601277.2015.1109405. 68

Tanner, D. (2012). Co-research with older people with dementia: Experience and reflections. *Journal of Mental Health*, 21(3): 296–306. DOI: 10.3109/09638237.2011.651658. 72, 75, 79, 83, 133

Tchalla ,A.E., Lachal, F., Cardinaud, N., Saulnier, I., Rialle, V., Preux, P.-M., and Dantoine, T. (2013) Preventing and managing indoor falls with home-based technologies in mild and moderate Alzheimer's Disease patients: Pilot study in a community dwelling. *Dementia and Geriatric Cognitive Disorders*. 36(3-4):251-261. DOI: 10.1159/000351863. 15

Teipel, S., Babiloni, C., Hoey, J., Kaye, J., Kirste, T., and Burmeister, O. K. (2016). Information and communication technology solutions for outdoor navigation in dementia. *Alzheimer's and Dementia*, 12(6): 695–707. DOI: 10.1016/j.jalz.2015.11.003. 46

The AAL programme (n.d.). [Website]: http://www.aal-europe.eu/. 3

The Alzheimer's Store (2020). [Website]: https://www.alzstore.com/clocks-alzheimers-dementia-s/1553.htm. 39

The Campaign to End Loneliness (n.d.). [Website]. The facts on loneliness: https://www.campaigntoendloneliness.org/the-facts-on-loneliness/. 57

The Lindsay Advocate (2019, July 4). Seniors' play park in Fenelon Falls: groundbreaking ceremony. *The Lindsay Advocate*. [Newspaper]: https://lindsayadvocate.ca/seniors-play-park-in-fenelon-falls-groundbreaking-ceremony/. 55

The National Alliance for Caregiving and AARP Public Policy Institute (2015, June). Caregiving in the U.S. [Report]: https://www.aarp.org/content/dam/aarp/ppi/2015/caregiving-in-the-united-states-2015-report-revised.pdf. 44

The National Seniors Council (2014). Report on the social isolation of seniors: 2013–2014: https://tinyurl.com/yycfu7kr. 60

UK Public General Acts (2005). Mental Capacity Act 2005 c 9. legislation.gov.uk: http://www.legislation.gov.uk/ukpga/2005/9/contents. 87

Unbehaun, D., Aal, K., Vaziri, D. D., Wieching, R., Tolmie, P., and Wulf, V. (2018, November) Facilitating collaboration and social experiences with videogames in dementia: results and implications from a participatory design study. *Proceedings of the ACM on Human-Computer Interaction 2, CSCW*, Article no. 175. DOI: 10.1145/3274444. 60

Unbehaun, D., Aal, K., Vaziri, D. D., Tolmie, P., Wieching, R., Randall, D., and Wulf, V. (2020). Social technology appropriation in dementia: investigating the role of caregivers in engaging people with dementia with a videogame-based training system. In *CHI '20: Proceedings of the 2020 CHI Conference on Human Factors in Computing Systems*, April 2020, pp. 1–15. DOI: 10.1145/3313831.3376648.

United Nations Sustainable Goals (n.d.). About the sustainable goals. [Website]: https://www.un.org/sustainabledevelopment/sustainable-development-goals/. 102, 138

United Nations. Department of Economic and Social Affairs, Population Division (2015). World Population Ageing 2015—Highlights (ST/ESA/SER.A/368). New York: United Nations, Department of Economic and Social Affairs, Population Division: https://www.un-.org/en/development/desa/population/publications/pdf/ageing/WPA2015_Highlights.pdf. 102

United Nations. Department of Economic and Social Affairs Ageing (2017, July 21). Ageing, older persons and the 2030 agenda for sustainable development. [Website]: https://www.un.org/development/desa/ageing/news/2017/07/ageing-older-persons-and-the-2030-agenda-for-sustainable-development.

United Nations (2018). Ageing. [Website]: http://www.un.org/en/sections/issues-depth/ageing/.

Uzor, S. and Baillie, L. (2014). Investigating the long-term use of exergames in the home with elderly fallers. *Proceedings of the SIGCHI Conference on Human Factors in Computing Systems.* DOI: 10.1145/2556288.2557160. 53

Uzor, S. and Baillie, L. (2019). Recov-R: evaluation of a home-based tailored exergame system to reduce fall risk in seniors. *ACM Transactions on Computer-Human Interaction (TOCHI),* 26(4): 1–38. DOI: 10.1145/3325280.

Van Bavel, J. J., Baicker, K., Boggio, P. S., Capraro, V., Cichocka, A., Cikara, M., Crockett, M. J., Crum, A. J., Douglas, K. M., Druckman, J. N., Drury, J., Dube, O., Ellemers, N., Finkel, E. J., Fowler, J. H., Gelfand, M., Han, S., Haslam, S. A., Jetten, J., Kitayama, S., Mobbs, D., Napper, L. E., Packer, D. J., Pennycock, G., Peters, E., Petty, R. E., Rand, D. G., Reicher, S. D., Schnall, S., Shariff, A., Skitka, L. J., Smith, S. S., Sunstein, C. R., Tabri, N., Tucker, J. A., Van der Linden, S., Van Lange, P., Weeden, K. A., Wohl, M. J. A., Zaki, J., Zion, S. R., and Robb, W. (2020). Using social and behavioural science to support COVID-19 pandemic response. *Nature Human Behaviour,* 4: 460–471: https://www.nature.com/articles/s41562-020-0884-z. DOI: 10.31234/osf.io/y38m9. 128

Van der Roest, H. G., Wenborn, J., Pastink, C., Dröes, R. M., and Orrell, M. (2017). Assistive technology for memory support in dementia. *Cochrane Database of Systematic Reviews,* Issue 6. Article no. CD009627. DOI: 10.1002/14651858.CD009627.pub2. 15

Vaportzis, E., Clausen, M. G., and Gow, A. J. (2017). Older adults perceptions of technology and barriers to interacting with tablet computers: a focus group study. *Frontiers in Psychology,* 8:1687. DOI: 10.3389/fpsyg.2017.01687; https://www.ncbi.nlm.nih.gov/pmc/articles/PMC5649151/. 58

Vaughn, J., Shaw, R. J., and Molloy, M. A. (2015). A telehealth case study. *Journal of the American Psychiatric Nurses Association,* 21(6): 431–432. DOI: 10.1177/1078390315617037. 17, 137

Vaziri, D. D., Aal, K., Ogonowski, C., Von Rekowski, T., Kroll, M., Marston, H. R., Poveda, R., Gschwind, Y. J., Delbaere, K., Wieching, R., and Wulf, V. (2016). Exploring user experience and technology acceptance for a fall prevention system: results from a randomized clinical trial and a living lab. *European Review of Aging and Physical Activity,* 13. DOI: 10.1186/s11556-016-0165-z.

Vaziri, D. D., Anslinger, M., Unbehaun, D., Wieching, R., Randall, D., Schreiber, D., and Wulf, V. (2019). Mobile health platforms for active and healthy ageing support in older adults—Design ideas from a participatory design study. *International Reports on Socio Informatics,* 16(2). DOI: 10.25819/ubsi/217.

Voegtlin, C. and Scherer, A. G. (2017). Responsible innovation and the innovation of responsibility: Governing sustainable development in a globalized world. *Journal of Business Ethics*, 143(2): 227–243. DOI: 10.1007/s10551-015-2769-z. 1

Vital Aging NetworkTM. (2020). Vital Aging. [Website]: https://vital-aging-network.org/.

Wada, M., Sixsmith, J., Harwood, G., Cosco, T. D., Fang, M. L., and Sixsmith, A. (2020). A protocol for co-creating research project lay summaries with stakeholders: guideline development for Canada's AGE-WELL network. *Research Involvement and Engagement*, 6(1): 1–8. https://researchinvolvement.biomedcentral.com/articles/10.1186/s40900-020-00197-3. DOI: 10.1186/s40900-020-00197-3. 116

Wan, L., Müller, C., Randall, D., and Wulf, V. (2016). Design of A GPS monitoring system for dementia care and its challenges in academia-industry project. ACM Transactions on Computer-Human Interaction (TOCHI), 23(5), 1-36. DOI: 10.1145/2963095. 37

Wang, X. J., Xu, W., Li, J. Q., Cao, X. P., Tan, L., and Yu, J. T. (2019). Early-life risk factors for dementia and cognitive impairment in later life: a systematic review and meta-analysis. *Journal of Alzheimer's Disease*, 67(1): 221–229. DOI: 10.3233/JAD-180856. 12

Ward, R. and Campbell, S. (2013). Mixing methods to explore appearance in dementia care. *Dementia*, 12(3): 337–347. DOI: 10.1177/1471301213477412. 79

Waycott, J., Vetere, F., Pedell, S., Kulik, H., Ozanne, E., Gruner, A., and Downs, J. (2013). Older adults as content producers. In Mackay, W. E., Brewster, S., and Bødker, S. (Eds.), In *CHI '13: Proceedings of the SIGCHI Conference on Human Factors in Computing Systems*. April 2013. Paris: ACM Press, pp. 39–48. DOI: 10.1145/2470654.2470662. 61, 134

Welsh, D., Morrissey, K., Foley, S., McNaney, R., Salis, C., McCarthy, J., and Vines, J. (2018). Ticket to talk: Supporting conversation between young people and people with dementia through digital media. In *Proceedings of the 2018 CHI Conference on Human Factors in Computing Systems – CHI '18*. Presented at the 2018 CHI Conference, ACM Press, Montreal, Quebec, Canada, pp. 1–14. DOI: 10.1145/3173574.3173949. 17, 60

West, E., Stuckelberger, A., Pautex, S., Staaks, J., and Gysels, M. (2017). Operationalising ethical challenges in dementia research—A systematic review of current evidence. *Age and Ageing*, 46(4): 678–687. DOI: 10.1093/ageing/afw250. 87

Williams, K., Arthur, A., Niedens, M., Moushey, L., and Hutfles, L. (2013). In-home monitoring support for dementia caregivers: A feasibility study. *Clinical Nursing Research*, 22(2): 139–150. DOI: 10.1177/1054773812460545. 16

Woolrych, R., Sixsmith, J., Lawthom, R., Makita, M., Fisher, J., and Murray, M. (2019). Constructing and negotiating social participation in old age: experiences of older adults living in urban environments in the United Kingdom. *Ageing and Society*. DOI: 10.1017/ S0144686X19001569. 43, 57, 137

Wootton, R. and Hebert, M. A. (2001). What constitutes success in telehealth? *Journal of Telemedicine and Telecare*, 7(Suppl 2): 3–7. DOI: 10.1017/S0144686X19001569. 109

World Health Organization (2007). Global age-friendly cities: a guide. Geneva: World Health Organization: https://www.who.int/ageing/publications/Global_age_friendly_cities_ Guide_English.pdf. 43

World Health Organization (2017a). Dementia: https://www.who.int/news-room/fact-sheets/ detail/dementia. 9

World Health Organization (WHO) (2017b). Global strategy and action plan on ageing and health: https://www.who.int/ageing/WHO-GSAP-2017.pdf?ua=1; 2017. Licence: CC BY-NC-SA 3.0 IGO. Geneva: World Health Organization. 9

WHO (2017c). Mental health: Global Dementia Observatory (GDO). https://www.who.int/mental_health/neurology/dementia/Global_Observatory/en/. 106

WHO (2019, September 19). Dementia: https://www.who.int/news-room/fact-sheets/detail/dementia. 10

WHO International (2020). 10 Priorities for a Decade of Action on Healthy Ageing: http://www. who.int/ageing/10-priorities/en/. 100

Wray, L. O., Shulan, M. D., Toseland, R. W., Freeman, K. E., Vásquez, B. E., and Gao, J. (2010). The effect of telephone support groups on costs of care for veterans with dementia. *Gerontologist*, 50(5): 623–631. DOI: 10.1093/geront/gnq040. 110

Yang, Y., Hirdes, J. P., Dubin, J. A., and Lee, J. (2019). Fall risk classification in community-dwelling older adults using a smart wrist-worn device and the resident assessment instrument-home care: prospective observational study. *JMIR Aging*, 2(1): e12153. DOI: 10.2196/12153. 30

Zarit, S. H. (2012). Positive aspects of caregiving: More than looking on the bright side. *Aging and Mental Health*, 16(6): 673–674. DOI: 10.1080/13607863.2012.692768. 127

Zhang, F. and Kaufman, D. (2015). The impacts of social interactions in MMORPGs on older adults' social capital. *Computers in Human Behavior*, 51, Part A: 495–503. DOI: 10.1016/j. chb.2015.05.034. 61

Zhang, F., Kaufman, D., Schell, R., Salgado, G., Seah, E. T. W., and Jeremic, J. (2017). Situated learning through intergenerational play between older adults and undergraduates. *Inter-*

national Journal of Educational Technology in Higher Education, 14, Article no. 16. DOI: 10.1186/s41239-017-0055-0. 61

Authors' Biographies

Andrew Sixsmith, Ph.D. is the joint Scientific Director of AGE-WELL Network of Centres of Excellence (NCE), the Director of Simon Fraser University's (SFU) Science and Technology for Aging Research (STAR) Institute, and a professor in the Department of Gerontology at SFU. He is past President of the International Society of Gerontechnology and was previously Director of the Gerontology Research Centre and Deputy Director of the Interdisciplinary Research in the Mathematical and Computational Sciences (IR-MACS) Centre at SFU. His research interests include technology for independent living, theories and methods in aging, and understanding the innovation process. His work has involved him in a leadership and advisory role in numerous major international research projects and initiatives with academic, government, and industry partners. He received his doctorate from the University of London and was previously a lecturer at the University of Liverpool at the Institute of Human Ageing and Department of Primary Care.

Judith Sixsmith, Ph.D. is a Professor in the School of Health Sciences at the University of Dundee. Her research interests reside in the areas of public health and social care where she explores the ways in which people living in disadvantaged communities experience processes of marginalization, taking care to include the voices of people who are seldom heard. She has expertise in qualitative methods including visual and textual designs and analyses. Often working within collaborative, gendered, participatory, and transdisciplinary approaches, Judith has directed several research projects on issues of healthy aging, dementia, place-making, and palliative and end-of-life care.

Mei Lan Fang, Ph.D. is an Assistant Professor in the School of Health Sciences at the University of Dundee. With a background in public health, specializing in social and health inequities, she has applied this lens in her research working across health disciplines and sectors as a transdisciplinary scientist and health sciences methodologist. Over the past ten years, across academic institutions in both Canada and in the United Kingdom, Mei has developed theory, methods, and practice in health-related areas of critical public health, ethnic and migration studies, environmental gerontology, aging and technology, global health promotion, and mental health and addiction.

Becky Horst, M.Sc. is a current Cognitive Neuroscience Ph.D. student at Western University. With a background in kinesiology and neuroanatomy, her thesis work largely revolves around projects that integrate exercise psychology, neuroanatomy, and cognition. Her research specifically focuses on older adults' perceptions of their cognitive abilities and the influence modifying their physical and cognitive health has on psychological perception of self, neurocognitive networks, and overall brain health. Becky has also been involved in developing Knowledge Translation materials for the AGE-WELL Network, contributing to the production of "The Future of Technology and Aging Research in Canada," as well as the STAR Institute's "Key Issues in Aging in the 21st Century" digital booklet.

Printed in the United States
by Baker & Taylor Publisher Services